Science Fiction and Futurism

CRITICAL EXPLORATIONS IN SCIENCE FICTION AND FANTASY
(a series edited by Donald E. Palumbo and C.W. Sullivan III)

1 *Worlds Apart? Dualism and Transgression in Contemporary Female Dystopias* (Dunja M. Mohr, 2005)

2 *Tolkien and Shakespeare: Essays on Shared Themes and Language* (ed. Janet Brennan Croft, 2007)

3 *Culture, Identities and Technology in the Star Wars Films: Essays on the Two Trilogies* (ed. Carl Silvio, Tony M. Vinci, 2007)

4 *The Influence of Star Trek on Television, Film and Culture* (ed. Lincoln Geraghty, 2008)

5 *Hugo Gernsback and the Century of Science Fiction* (Gary Westfahl, 2007)

6 *One Earth, One People: The Mythopoeic Fantasy Series of Ursula K. Le Guin, Lloyd Alexander, Madeleine L'Engle and Orson Scott Card* (Marek Oziewicz, 2008)

7 *The Evolution of Tolkien's Mythology: A Study of the History of Middle-earth* (Elizabeth A. Whittingham, 2008)

8 *H. Beam Piper: A Biography* (John F. Carr, 2008)

9 *Dreams and Nightmares: Science and Technology in Myth and Fiction* (Mordecai Roshwald, 2008)

10 *Lilith in a New Light: Essays on the George MacDonald Fantasy Novel* (ed. Lucas H. Harriman, 2008)

11 *Feminist Narrative and the Supernatural: The Function of Fantastic Devices in Seven Recent Novels* (Katherine J. Weese, 2008)

12 *The Science of Fiction and the Fiction of Science: Collected Essays on SF Storytelling and the Gnostic Imagination* (Frank McConnell, ed. Gary Westfahl, 2009)

13 *Kim Stanley Robinson Maps the Unimaginable: Critical Essays* (ed. William J. Burling, 2009)

14 *The Inter-Galactic Playground: A Critical Study of Children's and Teens' Science Fiction* (Farah Mendlesohn, 2009)

15 *Science Fiction from Québec: A Postcolonial Study* (Amy J. Ransom, 2009)

16 *Science Fiction and the Two Cultures: Essays on Bridging the Gap Between the Sciences and the Humanities* (ed. Gary Westfahl, George Slusser, 2009)

17 *Stephen R. Donaldson and the Modern Epic Vision: A Critical Study of the "Chronicles of Thomas Covenant" Novels* (Christine Barkley, 2009)

18 *Ursula K. Le Guin's Journey to Post-Feminism* (Amy M. Clarke, 2010)

19 *Portals of Power: Magical Agency and Transformation in Literary Fantasy* (Lori M. Campbell, 2010)

20 *The Animal Fable in Science Fiction and Fantasy* (Bruce Shaw, 2010)

21 *Illuminating* Torchwood: *Essays on Narrative, Character and Sexuality in the BBC Series* (ed. Andrew Ireland, 2010)

22 *Comics as a Nexus of Cultures: Essays on the Interplay of Media, Disciplines and International Perspectives* (ed. Mark Berninger, Jochen Ecke, Gideon Haberkorn, 2010)

23 *The Anatomy of Utopia: Narration, Estrangement and Ambiguity in More, Wells, Huxley and Clarke* (Károly Pintér, 2010)

24 *The Anticipation Novelists of 1950s French Science Fiction: Stepchildren of Voltaire* (Bradford Lyau, 2010)

25 *The Twilight* Mystique: *Critical Essays on the Novels and Films* (ed. Amy M. Clarke, Marijane Osborn, 2010)

26 *The Mythic Fantasy of Robert Holdstock: Critical Essays on the Fiction* (ed. Donald E. Morse, Kálmán Matolcsy, 2011)

27 *Science Fiction and the Prediction of the Future: Essays on Foresight and Fallacy* (ed. Gary Westfahl, Wong Kin Yuen, Amy Kit-sze Chan, 2011)

28 *Apocalypse in Australian Fiction and Film: A Critical Study* (Roslyn Weaver, 2011)

29 *British Science Fiction Film and Television: Critical Essays* (ed. Tobias Hochscherf, James Leggott, 2011)

30 *Cult Telefantasy Series: A Critical Analysis of* The Prisoner, Twin Peaks, The X-Files, Buffy the Vampire Slayer, Lost, Heroes, Doctor Who *and* Star Trek (Sue Short, 2011)

31 *The Postnational Fantasy: Essays on Postcolonialism, Cosmopolitics and Science Fiction* (ed. Masood Ashraf Raja, Jason W. Ellis and Swaralipi Nandi, 2011)

32 *Heinlein's Juvenile Novels: A Cultural Dictionary* (C.W. Sullivan III, 2011)

33 *Welsh Mythology and Folklore in Popular Culture: Essays on Adaptations in Literature, Film, Television and Digital Media* (ed. Audrey L. Becker and Kristin Noone, 2011)

34 *I See You: The Shifting Paradigms of James Cameron's* Avatar (Ellen Grabiner, 2012)

35 *Of Bread, Blood and* The Hunger Games: *Critical Essays on the Suzanne Collins Trilogy* (ed. Mary F. Pharr and Leisa A. Clark, 2012)

36 *The Sex Is Out of This World: Essays on the Carnal Side of Science Fiction* (ed. Sherry Ginn and Michael G. Cornelius, 2012)

37 *Lois McMaster Bujold: Essays on a Modern Master of Science Fiction and Fantasy* (ed. Janet Brennan Croft, 2013)

38 *Girls Transforming: Invisibility and Age-Shifting in Children's Fantasy Fiction Since the 1970s* (Sanna Lehtonen, 2013)

39 Doctor Who *in Time and Space: Essays on Themes, Characters, History and Fandom, 1963–2012* (ed. Gillian I. Leitch, 2013)

40 *The Worlds of* Farscape: *Essays on the Groundbreaking Television Series* (ed. Sherry Ginn, 2013)

41 *Orbiting Ray Bradbury's Mars: Biographical, Anthropological, Literary, Scientific and Other Perspectives* (ed. Gloria McMillan, 2013)

42 *The Heritage of Heinlein: A Critical Reading of the Fiction Television Series* (Thomas D. Clareson and Joe Sanders, 2014)

43 *The Past That Might Have Been, the Future That May Come: Women Writing Fantastic Fiction, 1960s to the Present* (Lauren J. Lacey, 2014)

44 *Environments in Science Fiction: Essays on Alternative Spaces* (ed. Susan M. Bernardo, 2014)

45 *Discworld and the Disciplines: Critical Approaches to the Terry Pratchett Works* (ed. Anne Hiebert Alton and William C. Spruiell, 2014)

46 *Nature and the Numinous in Mythopoeic Fantasy Literature* (Christopher Straw Brawley, 2014)

47 *J.R.R. Tolkien, Robert E. Howard and the Birth of Modern Fantasy* (Deke Parsons, 2014)

48 *The Monomyth in American Science Fiction Films: 28 Visions of the Hero's Journey* (Donald E. Palumbo, 2014)

49 *The Fantastic in Holocaust Literature and Film: Critical Perspectives* (ed. Judith B. Kerman and John Edgar Browning, 2014)

50 Star Wars *in the Public Square:* The Clone Wars *as Political Dialogue* (Derek R. Sweet, 2016)

51 *An Asimov Companion: Characters, Places and Terms in the Robot/Empire/Foundation Metaseries* (Donald E. Palumbo, 2016)

52 *Michael Moorcock: Fiction, Fantasy and the World's Pain* (Mark Scroggins, 2016)

53 *The Last Midnight: Essays on Apocalyptic Narratives in Millennial Media* (ed. Leisa A. Clark, Amanda Firestone and Mary F. Pharr, 2016)

54 *The Science Fiction Mythmakers: Religion, Science and Philosophy in Wells, Clarke, Dick and Herbert* (Jennifer Simkins, 2016)

55 *Gender and the Quest in British Science Fiction Television: An Analysis of* Doctor Who, Blake's 7, Red Dwarf *and* Torchwood (Tom Powers, 2016)

56 *Saving the World Through Science Fiction: James Gunn, Writer, Teacher and Scholar* (Michael R. Page, 2017)

57 *Wells Meets Deleuze: The Scientific Romances Reconsidered* (Michael Starr, 2017)

58 *Science Fiction and Futurism: Their Terms and Ideas* (Ace G. Pilkington, 2017)

Science Fiction and Futurism
Their Terms and Ideas

ACE G. PILKINGTON

Foreword by David Brin

CRITICAL EXPLORATIONS IN
SCIENCE FICTION AND FANTASY, 58
Series Editors Donald E. Palumbo and C.W. Sullivan III

McFarland & Company, Inc., Publishers
Jefferson, North Carolina

LIBRARY OF CONGRESS CATALOGUING-IN-PUBLICATION DATA

Names: Pilkington, Ace G., author. | Brin, David, writer of foreword.
Title: Science fiction and futurism : their terms and ideas / Ace G. Pilkington ; foreword by David Brin.
Description: Jefferson, North Carolina : McFarland & Company, Inc., Publishers, 2017. | Series: Critical explorations in science fiction and fantasy ; 58 | Includes bibliographical references and index.
Identifiers: LCCN 2017006486 | ISBN 9780786498567 (softcover : acid free paper) ∞
Subjects: LCSH: Science fiction—Terminology. | Futurism (Literary movement) | Science fiction—Dictionaries. | Science fiction—Encyclopedias. | Science fiction—History and criticism. | Literary form—Terminology.
Classification: LCC PN3433.4 .P55 2017 | DDC 809.3/876203—dc23
LC record available at https://lccn.loc.gov/2017006486

BRITISH LIBRARY CATALOGUING DATA ARE AVAILABLE

ISBN (print) 978-0-7864-9856-7
ISBN (ebook) 978-1-4766-2955-1

© 2017 Ace G. Pilkington. All rights reserved

No part of this book may be reproduced or transmitted in any form or by any means, electronic or mechanical, including photocopying or recording, or by any information storage and retrieval system, without permission in writing from the publisher.

Front cover image of faces © 2017 harshvardhanroy/iStock

Printed in the United States of America

McFarland & Company, Inc., Publishers
Box 611, Jefferson, North Carolina 28640
www.mcfarlandpub.com

To the memory of Charles Sheffield,
who encouraged an earlier version of this book,
when he could have written it so much better himself

It is worth asking how much robots trouble us because of differences that trigger old fears and how much they trouble us because of similarities that raise new questions. —the Author

Table of Contents

Acknowledgments xi
Foreword by David Brin 1
Key and Abbreviations 3
Preface 5
Introduction 11

Part One: The Terms of Science and Its Fictions

Alter Ego	21	Cyberspace	55
Anachronism	24	Death Ray	58
Artificial Intelligence	27	Deep Blue	60
Asimov's Three Laws of Robotics	29	Doomsday Machine	66
		Fermis	71
Atomics	34	First Contact	74
Bad Code	41	Frankenstein Complex	77
BEM (Bug-Eyed Monster)	42	Gas Giant	80
Big Dumb Objects	43	Goldilocks Planet	81
Bobble	45	"He's dead, Jim."	81
Chronotransference	45	Internet of Things	82
Clarke's Laws	46	Ion Drive	83
Corpsicle	47	Juno Spacecraft	86
Cosmogony of the Future	48	Kludge	89
Cryonics	51	MacGyver	91

Metalaw	92	Solar Sail	126
Multiplex Parenting	97	Solo Parenting	130
Mutant	99	Stepford Wives	131
Nanotechnology	105	Teleportation	132
Neural Lace	108	Time Bunny	136
Pellegrino, Powell and Asimov's Three Laws of Alien Behavior	111	Time Machine	137
		Timequake	139
		Timeskip	140
Positronic Brain	113	Timeslip	140
Railgun	114	Uncanny Valley	142
Realtime	118	Uplift	145
Robot	119	Uterine Replicator	149
Robotics	123	Utility Fog	151
SETI	123	Wolfling	152

Part Two: Genre Terms

Alternate (or Alternative) History	155	Proto–Science Fiction	178
		Rim World	178
Archetype	156	Ruritania	179
Cassandras	157	Science Fantasy	180
Fantasy	160	Science Fiction	181
Faust	161	Scientific Romance	182
Fictions of Nuclear Disaster	166	Sci-Fi	184
Fixup	170	Separable Soul	185
Hard SF	170	Speculative Fiction	185
Lost Colony	171	Speculative History	188
Lost World Story	172	Speculative Nonfiction	189
Monomyth	173	Technothriller	190

Works Cited 193

Index 213

Acknowledgments

My thanks to the many kind, intelligent, and extremely well-informed people who made this book possible. To the extraordinary Charles Sheffield, who said nice things about an early (unpublished) version of the book and mentioned me in one of his—*The Borderlands of Science*. To David Brin, who, in addition to being the perfect example of a scientist, futurist and science fiction author, took time from his very busy schedule to write a wonder-filled Foreword. To Nancy Kress, who had encouraging words for the project. To Matthew Wilhelm Kapell, who insisted that it was time for me to write this book and provided support along the way. To Donald Palumbo, who is a remarkable editor—erudite, understanding and sympathetic, and without whom this project would not have gone forward. To the people at McFarland for great suggestions and greater patience. To the students in my science fiction and futurism classes who read and argued over these entries. To Elaine P. Pearce for dedicated proofreading and cheerful encouragement. To my wife, Olga, for every possible kind of help, including clever editing and a transcendent ability to ameliorate all the problems that come along with writing. And to everyone whose work I mentioned, described or quoted, my thanks for making the world a smarter and far more entertaining place.

Foreword by David Brin

One of the foremost traits of Speculative or Science Fiction is the genre's resistance to prim definition. If you cram 500 cubic meters of SF authors, critics and fans into a 300 cubic meter room, you won't get an implosion or explosion of organic matter. Your output will be a tsunami—a supernova—of opinions ... not the least of them about the meaning of "science fiction." Of course, this led to many sagacious tomes, from *The Encyclopedia of Science Fiction* to Hartwell and Cramer's *The Ascent of Wonder*, to Brian Aldiss's *Trillion Year Spree*. Like a crowd of proverbial blind men groping at elephants, the scholars of our impudent field not only come up with a variety of anatomical comparisons, but also rather piqued—sometimes even angry—pachyderms.

"Science fiction" was always an iffy term. My own view of the genre has little or nothing to do with *science* at all. What name would I prefer? *Speculative History!* Think about it. Only ten percent or so of science fiction authors are scientifically trained. But all of us are complete history junkies. We pore over the tragic, poignant past—the epic of stumbling advances and brilliant calamities. We speculate what history might have been like if this or that happened or if the rules had been different. Above all, how might we *extrapolate* this gorgeous litany of stupid, horrific or well-meaning errors that is called the human experience? No, the trait that is most shared among high quality SF stories, novels and films is a willingness to contemplate *change*. That things might be different for our grandchildren than they are for us—better or worse. And that the human coping process is itself one of our most fascinating traits, proper grist for riveting literature.

Which brings us to this volume. It lured you in by offering alphabetized topics, implying you'll be offered prim definitions, or encyclopedic references for a wide array of bold notions, some of them coming true right now, before our very eyes. Some that are clearly impossible. All of them introduced to millions and conveyed dramatically by this rambunctious art form—sci fi.

But no, this is not another SF encyclopedia. Ace Pilkington here takes on everything from Anachronism to Alter Egos to First Contact from the perspective of someone a lot like *you* ... logical, committed to thoroughness, scarily well-read, but above all opinionated and fiercely eager to argue about ideas! Even ... especially ... concepts for which there are no living or palpable examples, from Bug Eyed Monsters to Teleportation. Things that have never actually been seen (and might never be in-the-future), but that have been chewed over from every angle, by authors and other creators, all of *them* logical, thorough, well-read, opinionated and fiercely argumentative! Comprehensive? This ain't. Provocative and enticing? Filled with "huh!" moments and leads to great stories? That describes this volume. And I hereby give you permission to scribble in the margins all the things the compiler left out.

Welcome to the cult ritual of this peculiar tribe. Arguing over what-might-be. Probing the horizons of possibility. Never being satisfied.

David Brin is an astrophysicist whose international bestselling novels include The Postman, Earth *and* Existence. *His nonfiction book about the information age—*The Transparent Society*—won the Freedom of Speech Award of the American Library Association.*

Key and Abbreviations

Key

Words or terms in **bold** are Part One main entries.

Words or terms in ***bold italics*** indicate entries in Part Two, on genres and similar material helpful in understanding the subjects discussed.

Abbreviations Used

AI—Artificial Intelligence

ASI—Artificial Superintelligence

IMDb—Internet Movie Database

OED—*Oxford English Dictionary*

SF—Science Fiction

SFWA—Science Fiction and Fantasy Writers of America

STNG—*Star Trek: The Next Generation*

Preface

This is a book of words and ideas invented or modified by *science fiction* writers, futurists, and scientists or to put it another way, by people who spend a significant amount of their time thinking about the future, predicting what it may be like and suggesting ways to make it so. As a result, this is a book about ideas and how they changed the world. Next, of necessity, it's a book about history because those ideas grew and those changes occurred gradually in time. And finally, this is a book about the terms that collected, clarified and crystallized those ideas, sometimes showing them off, sometimes slowing them down, and sometimes propelling them to fame and making them the common currency of our culture. Sometimes the terms are just statements of fact, sometimes they are hopes for the future, desperate fears or even war cries. It's not unusual for such terms to be thought experiments in the minds of scientists or writers of science fiction. The words themselves may become battlegrounds. **Robot**, a term that was first used in Čapek's play R.U.R. (standing for Rossum's Universal Robots), was meant as a warning of terrible things to come. Isaac Asimov and other techno-optimists made it a part of the happy future they foresaw.

The genre of science fiction has for more than two centuries become ever more popular, powerful and pervasive and ever more closely connected to science and scientists. It might be called the language of the future, but it is also very much a part of the present, and to a surprising extent and for a variety of reasons, a significant component in the past. Indeed, so many aspects of our lives are entwined with this genre that it not only makes up much of what we read and watch, but has also become an important part of the way we think, and even to some extent the means by which we live. In addition, it provides patterns for the way we may live tomorrow. In *Make It So: Interaction Design Lessons from Science Fiction*, Nathan Shedroff and Christopher Noessel use science fiction in its various forms as both a source of new ideas and a testing ground for design possibilities in the real world. They say, "Whether we like it or not, the fictional technology seen in *sci-fi* sets audience expectations for what exciting things are coming next" (6). Another way to put it might be to say that science fiction helps to prepare us for what's about to happen. Arthur C. Clarke wrote, "A critical ... reading of science fiction is essential training for anyone wishing to look more than ten years ahead. The facts of the future can hardly be imagined *ab initio* by those who are unfamiliar with the fantasies of the past" (1971, xiii).

Such futurists may be science fiction writers or scientists or scholars from almost any discipline or they may belong, as Stan Davis suggests he does, to "that peculiar breed of thinker who anticipates the changes sweeping through society and identifies the business opportunities they present" (xii). Clearly, an interest that was once confined to science fiction writers and scientists has now become a preoccupation for much of society, and any list of those who have been paid to do the work of futurists would include people as different from each other as Herman Kahn (who helped to develop the Scenario Method of predicting, among other things, the future of nuclear weapons [Cornish 93]) and Ray Bradbury (who took time off from writing poetic fantasies to help design the Epcot Center and giant malls [Rose]).

Science, science fiction and futurism are generating enormous energy. They are now, and they have been for some time, changing the nature of our society, and equally important for the purposes of this book, they are changing the nature of our vocabulary, inventing new words and altering or adding to the meanings of old ones, reshaping how we speak and behind that, how we think. Perhaps nothing is more extraordinary in the world at the present moment than the ever-increasing speed of change and our attempts to understand, predict and control it.

I will have more to say about these issues in the Introduction, but for now I want to confine myself to what's in this book and how it works. Science fiction and futurism can be enormously important and extraordinarily entertaining, but they can also be confusing. The words they employ are often complex in and of themselves, and they frequently come with histories, some long, some short, that impact their meanings, plus ongoing arguments about what they *should* mean. In addition, there are disputes about how the things those words represent should be used now and in the future.

The purpose of this book is to make the complexities of those terms clear. Entries are designed to answer the questions: what is it, where did it come from, and what is it about to become? Though there is a scholarly component, including many quotations and examples and an extensive bibliography, the language throughout is straightforward, and the explanations begin at the beginning so that no background information is necessary. There are no mathematical formulas, and nothing in High Valyrian or Klingon. Reading any one of the entries will answer questions on a particular subject. Some such as **BEM (Bug-Eyed Monster)** and **Big Dumb Objects** are short. Others such as **Nanotechnology** and **Teleportation** are longer. A few like **Atomics** and **Mutant** are longer still because they deal with a combination of issues and more complex histories. **Deep Blue**, for example, takes both a practical and philosophical look at the question of increasing machine intelligence and whether or not we should create **Artificial Intelligence**s that are smarter— perhaps far smarter—than humans. *Faust*, an entry in Part Two, gives the long history of the struggle between those who wanted knowledge and those who wielded authority, beginning with magic and ending with science. The entries are connected by plentiful cross references, and one of the more interesting results of those references is the possibility of looking at all the terms connected to a single subject, even when it is not immediately obvious how closely related they are. For instance, the subject of time travel has generated an extraordinary range of thought experiments, fantasies and strange new terms, including *Alternate History*, **Bobble, Chronotransference, Cryonics, Real-time,** *Speculative History,* **Time Bunny, Time Machine, Timequake, Timeskip,** and **Timeslip**. Plus, the **Time Machine** entry includes a discussion of Block Time, the idea that past, present and future all exist at once. Sometimes even comparatively short entries, of necessity, cover a range of subjects. In addition to its main subject, **Juno Spacecraft** looks at the naming of names in the solar system, the increasing number of robotic spacecraft being sent out from Earth, and the connection between Legos and Roman gods. Reading all (or even most) of the entries will serve as a detailed introduction to and a guided tour of the vast territories occupied by science fiction and futurism.

This book is not comprehensive; given the enormous quantity of information that stands behind such terms, no single source of any kind or size could be, but the book is representative. I've used this vocabulary in science fiction and futurism courses, and Hugo and Nebula award winner Charles Sheffield used it as a supplement in a science fiction workshop. I expect the book itself will be employed in similar ways, but it is equally well suited for individual readers who want to make the journey on their own. To make the book more useful and the journey easier, I have included a Part Two: Genre Terms, with entries on genres and similar material that is peripheral to the book itself but still helpful in understanding the subjects that the book discusses. As the Key indicates, main entries appear in **bold** type as with **Railgun**. Entries in Part Two are indicated in ***bold italics***, such as ***Cassandras***.

The entries in this book are arranged alphabetically with, as I mentioned, many cross references. There is a table of contents that lists all the entries, and a detailed index. I have tried to get at the essence of the terms, to see the ideas behind them, to get perhaps a glimmer of the magic in their creation and to look forward to their future. To take a short example, Kevin Ashton coined **Internet of Things** in 1999. It's not in dictionaries yet, but it's commonly used to mean "the rapidly growing network of everyday objects that have been equipped with sensors, small power supplies, and internet addresses" (Howard xix). Ashton's idea, though, was much bigger. He says, "We need to empower computers with their own means of gathering information, so they can see, hear and smell the world for themselves, in all its random glory." Ashton's idea was to get beyond "data originated by people," to get beyond human bias and human error to a kind of data that would be more accurate, more plentiful and, for humans at least, much less work. Ashton's idea and his term represent a step toward a smarter, better, more helpful world.

The same may be said for **MacGyver**, which in my definition is a verb meaning to fix, adjust or make something using whatever items are at hand, to improvise a solution to a problem using a minimum of material and a maximum of scientific knowledge. Often, though, there is a perhaps unjustified feeling that the solution is substandard. Several years ago when I was recovering from major surgery, a nurse who was having trouble changing my complicated surgical dressing said to me, "I'm going to have to MacGyver this." I found the prospect unsettling. In *Abundance: The Future Is Better Than You Think*, futurists Peter H. Diamandis and Steven Kotler hold MacGyver up as a model, "Ask yourself that fabled DIY question: What would MacGyver do? Well, MacGyver would empty his pockets and get the job done with a roll of Scotch tape, a piece of paper napkin, and a ball of spit—which, as it turns out, is exactly the solution we need" (193). They then go on to explain the principles behind setting up diagnostic tools which cost practically nothing—and yes, Scotch tape "generated X-rays" (194).

In *Hacking Matter*, a book about "programmable matter," which may be the technological revolution that comes after the **Internet of Things**, **Nanotechnology** and **Utility Fog**, Wil McCarthy says, "We use the levers and pulleys of technology to shape our world, but what we really want is a world that obeys our spoken commands and reconfigures itself to our unvoiced wishes. What we really want—what we've always wanted—is magic. The future is where these two notions converge" (3). And so it goes, from science fiction to fantasy, from magic to science and back again, as scientists, futurists, and SF writers remake our world. Humans have always been good at turning dreams to reality, at planning and even predicting the future, and now we stand at a point where our new realities may transcend even our oldest, wildest imaginings. The hand holding a wand will not belong to a wizard but to a futurist. The words and ideas in these pages truly hold more power than magic spells, but they are much easier to explain, and it is possible to contain them in this one volume.

Introduction

Science and science fiction are twins, with no one sure which is the elder. As with many twins, they have similar but not identical interests, a common language which they invented and speak with ease but which puzzles outsiders, and the ability to inspire and encourage each other, sometimes with no more than a look or a word. They finish each other's sentences and think the same thoughts at almost the same time even when they're far apart. It may initially seem surprising that two things as firmly logical as science and as resolutely imaginative as science fiction would be so much alike. But they think about the same things, and often, they think about them in the same ways. As George Zarkadakis says in his book about **artificial intelligence**, *In Our Own Image,* "Our brains are hard-wired for telling, and for listening to, stories." He continues, "I have argued that narratives ultimately dictate our artistic and scientific endeavors throughout history. But this relationship is not one-directional.... Our discoveries feed into our stories, and vice versa. One cannot separate the two" (305). And indeed science and science fiction have become inseparable, bound together in part by their common stories, interconnected thought experiments, and shared language. This book is about that close sibling relationship and those all-but-magical words. It is also about how those twins, gradually at first and then ever more swiftly, took over the world.

There are almost as many different definitions of ***science fiction*** as there are definers, as the entry on the subject in this book makes clear. For now, though, I'm more concerned about the starting point of the collaboration between science and science fiction and what resulted from it. Isaac Asimov's definition is particularly relevant. He says (in *Asimov on Science Fiction*) that SF is about "events played against social backgrounds that do not exist today… " but "could, conceivably, be derived from our own by appropriate changes in the level of science and technology" (17). He goes on to argue "that the field can scarcely have existed in its true sense until the time came when the concept of social change through alterations in the level of science and technology had been evolved in the first place" (18). He sets that time at the point when "the rate of change ... becomes great enough to be detected in ... an individual lifetime" (18). He cites the Industrial Revolution as the watershed event and declares, "Science fiction had to be born sometime after 1800" (18).

Brian Stableford begins the Chronology of his *Historical Dictionary of Science Fiction Literature* in 1726, and cites for 1771, "*L'An 2240* by Louis-Sebastian Mercier, the best-selling book of its era in France" and "the first.... Utopian society situated in the future whose evolution has been enabled by technology" (xiii). Both Asimov and Stableford are looking for that point in time when it became clear to many people that science (and the technologies that came along with it) was changing the nature of society, and would continue to do so in ever more important ways and at an increasing rate of speed.

What was coming, indeed what had already begun, was, in the words of British historian J. M. Roberts, "The spread of the experimental method, the growth of the systematic investigation of nature, the coordination of knowledge drawn from many, many different fields." These things, not surprisingly, are reflected in literature, and not just in what we now call science fiction, but in utopias, fantasies, and even in the newly popular genre of literary fairy tales, all as a response to, an encouragement of, or a reaction against the transformations that science was making in the world. Here is what folklorist Jack Zipes says about literary fairy tales (which became popular in France just before the end of the seventeenth century), "The once upon a time is not a past designation but futuristic: the timelessness of the tale and its lack of geographical specificity endow it with utopian connotations—'utopia' in its original meaning designated 'no place,' a place that no one had ever envisaged." He concludes with a statement that might well be true for much of the new writing that was driven by science, "We form and keep the utopian kernal of the tale safe in our imaginations with hope" (xiii).

The most important thing about science fiction may well be that it is not primarily a genre but as Istvan Csicsery-Ronay said "a mode of awareness." Or in the words of legendary SF author and editor Frederick Pohl, "I think science fiction is more a state of mind than a particular category of literature." Ultimately, it is (and has been from the beginning) an encouragement for the new uses of science in the world and the changes that inevitably follow. It has been analyzed in various ways in different times. As Franz Rottensteiner says in his

Introduction to *The Black Mirror and Other Stories,* "The first studies of SF in Germany were undertaken by sociologists…. They tended to see a direct line from utopias to science fiction" (xii).

That direct line is easy to find and not just in Germany. Almost all national histories of science fiction include utopian works, and some come remarkably early, such as Louis-Sebastian Mercier's *L'An 2240* already mentioned and the anonymous 1804 Spanish work *Selenopolis.* Leland Fetzer in his *Pre-Revolutionary Russian Science Fiction* says, "In approximately the second quarter of the nineteenth century in many countries in both the New and Old Worlds, in response to the promptings of the age, a number of writers turned to the question of contemporary science and what it meant for society" ("Introduction" i). He mentions Mary Shelley, Poe and Jules Verne and then continues, "In response to the same unspoken imperative…. Russian writers also turned to the place of science in society, and particularly—since scientific discoveries clearly imply accelerated and extensive future change—to the compelling question of the shape of things to come, to the nature of future societies … to utopias"(i).

It is worth noting that Fetzer's earliest example, Faddei Bulgarin's *Plausible Fantasies or a Journey to the Twenty-Ninth Century,* was published in 1824, only six years after *Frankenstein* and a century before the genre-centric history of science fiction is supposed to begin. Bulgarin's predictions include (among other things) air travel, parachute troops, rockets as accurate weapons, automobiles, submarines, underwater crops, air guns, a condensed food source, "a little of which was sufficient to feed a whole family for at least a month" (15), plus machines to turn air into water and produce hydrogen as a fuel for cooking. Many of these ideas have been credited to Jules Verne, but this was a movement of many minds, responding to a great and nearly universal transformation, not the work of a single genius.

The impulse behind early works of science fiction could, of course, be served by other kinds of narratives and was not limited to utopias. Jules Verne's "extraordinary voyages" (fictional and nonfictional) were designed as a scientific exploration of the entire world, past, present and future, and it was a world in which, as he said in *The Castle of the Carpathians,* "Everything can happen…. If our story does not seem to be true today, it may seem so tomorrow thanks to the resources of science, which are the wealth of the future" (33). Magazines dedicated to such actual and armchair explorations sprang up during the last decades of the nineteenth century in America, France and Russia.

14   Introduction

Commenting on the Russian magazines *Nature and People* and *Around the World*, Anindita Banerjee says they "created the template for a new kind of Russian subject: armchair geographers, scientists, and explorers, familiar with the powers of science and technology and possessing a spatial consciousness that extended to the farthest reaches of the planet and beyond" (17, 19). This vision of the planet would be reinforced and validated once space travel and space photography became a reality. As Charles Sheffield wrote in *Earth Watch: A Survey of the World from Space*, "Only recently, coinciding with the first images of the earth as a whole ... have we come to recognize that earth must be regarded as one entity, a cohesive and finite life-support system for the creatures that inhabit it" (7).

There is no doubt that one of the driving impulses behind Verne's work was an explanation for the uses of science that presently existed and an extrapolation of the uses to come. As he said, "I merely use my imagination and literary skill to argue from what is possible to what may be possible—tomorrow" (Taves and Michaluk 53). Nor was Verne alone. There was a worldwide outpouring of fictions and manifestoes. Beginning in 1902, China's first science fiction magazines were established "in order to stimulate people's interest in science and technology" (Wu xiii). In 1894, the Russian journal of popular science *Nature and People* justified expanding its parameters by declaring, "Science and technology are defining modern reality by transforming not just everyday life, but the ways in which we think and imagine. A new kind of writing *nauchnaia fantastika*, scientific fantasy, is playing a not inconsequential role in this process. Is it not in the imagination where bold theories and amazing machines are first born?"(cited in Banerjee 1)

Nor was it only magazines with—in some sense—specialized interests that published science fiction. Periodicals such as *The Strand, Pearson's Magazine, McClure's* and others like them published various forms of science fiction and fantasy, and in some cases competed with each other for the most popular authors. Histories of science fiction often downplay the intimate connection between science and science fiction, and in this context it might be well to remember that one term for SF works in the 1890s was "'invention novels,' and in those tales was forecast a ... parade of aircraft, submarines, tanks, and robots" (Moskowitz 17).

Often, H. G. Wells is held up as the only important writer of SF during this time, but there was an unusual range of talent, including (to mention only the still-famous) Rudyard Kipling, Arthur Conan Doyle, H. Rider Haggard,

Sax Rohmer, Jack London, Jerome K. Jerome, and even Mark Twain. The London magazines also included stories from French and Russian authors. As A. Kingsley Russell writes, "H. G. Wells has been given the credit for being the founding father of science fiction, but in his day he had to compete with some highly accomplished writers, many of whom at the time possibly even outstripped him in terms of both sales and fame" (vii).

There was an age of magazine science fiction that was far earlier and much broader in its appeal than the one that is usually cited as the beginning of SF. More importantly for the argument I'm pursuing, there was an amazing number of ideas which would find their way into later science fiction and into science. For example, the awakening of Robert H. Goddard's interest in rockets and space is often attributed to Wells, but the version he gives in his autobiography is more complex, "In January, 1898, there appeared for several months in the *Boston Post* the story *Fighters from Mars or the War of the Worlds*.... This as well as the story that followed it, *Edison's Conquest of Mars*, by Garritt P. Serviss, gripped my imagination tremendously" (cited in Moskowitz 28). "Serviss was an astronomer and popular science writer," and this was his first SF story (28).

Clearly, there was in the new mass-market magazines, as there had been in one form or another from the beginning of the interactions between science and science fiction, a movement of many people and much talent. There were also surprisingly accurate predictions of the wonders to come, successful recruiting of new scientists, plus clear and cogent explanations of emerging scientific principles. A good example of all these things in one person is Garritt P. Serviss; he is an especially useful example because he was in some ways typical of his time, an ordinary writer whose ideas were sometimes extraordinary. He was not only one of the inspirations for Goddard, but he also gave Carnegie lectures on science, wrote popular books on astronomy and one of the first books on relativity, and in 1909, imagined "atomic energy liberated and harnessed to drive a rocket to the planet Venus" (Searles). In the words of A. Langley Searles, "His conception is uncannily close to truth; he names uranium as the raw material from which is extracted the vital substance, a 'crystallized powder' which releases its energy on proper treatment." In Serviss's words from *A Columbus of Space*, "Every one of those minute grains ... is packed with as much potential energy as that of a ton's weight suspended a mile above the earth." *Edison's Conquest of Mars* is equally prescient: "It is the first to portray a battle fought by space craft in the airless void; and possibly the first also to propose the use of sealed suits that enable men to traverse a vacuum" (Searles).

The relationship between science fiction predictions and the inventions of science has always ranged somewhere between startling prophecies and inspired (and inspirational) plans. Among his many predictions, Jules Verne foresaw that even when a large storm was raging on the surface of the ocean, the depths would remain calm, a prediction that Simon Lake, inventor of "the first submarine to operate successfully in the open sea" (Miller and Walter x), verified twenty-five years later, saving himself and his vessel because he had read and remembered Verne's *Twenty Thousand Leagues Under the Sea*. Lake called Verne "the director-general of my life" (cited in Miller and Walter x). Verne was the first to suggest (wrongly) that it was possible to cross the South Pole by sailing under it (ix), and the U.S. nuclear submarine which was the first to execute such a maneuver at the North Pole was named the *Nautilus* to honor (among others) Jules Verne and his fictional submarine (ix). "Admiral Richard Byrd, on his way to the South Pole," also called Verne his "main inspiration" (x).

While direct communication between science fiction writers and the scientists or inventors they inspire is not always possible, in 1898, Jules Verne sent a cable to Simon Lake which said, among other things, "While my book 'Twenty Thousand Leagues Under the Sea' is entirely a work of imagination, my conviction is that all I said in it will come to pass. A thousand mile voyage in the Baltimore submarine boat (The Argonaut) is evidence of this. This conspicuous success of submarine navigation in the United States will push on under-water navigation all over the world" (Verne "Cable").

However, Verne's most remarkable predictions (which led to communications between his grandson and Frank Borman) were not concerned with air, land, or even the sea, but with space itself in *From the Earth to the Moon* (1865), and *All Around the Moon* (1870). In his biography of his grandfather, Jean Jules-Verne says, "Verne's eminently *scientific* [emphasis his] fiction never received a finer consecration.... 'It cannot be a mere matter of coincidence,' Frank Borman, the astronaut wrote to me in 1969. 'Our space vehicle [Apollo 8] was launched from Florida, like Barbicane's; it had the same weight and the same height, and it splashed down in the Pacific a mere two and a half miles from the point mentioned in the novel'" (93). "Verne's capsule" has the same "cylindro-conical shape" (Taves and Michaluk 2) as well. He correctly calculated escape velocity (12,000 yards per second) and a typical flight time

Introduction   17

(97 hours and 20 minutes) and suggested the use of rockets as retrojets, a suggestion picked up by Russian and German rocket scientists. Nor is this all. "Verne's astronauts experience weightlessness; they orbit at the same distance above the moon" (Taves and Michaluk 2). In addition, he saw "with uncanny accuracy the motivations for 'the space race' of the 1950s and 1960s" (Taves and Michaluk 2).

Charles Sheffield writes, "We know for a fact that Tsiolkovsky, the father of the Russian space program, was inspired by Verne. Hermann Oberth, whose work in turn inspired Wernher von Braun, discovered Verne's *From the Earth to the Moon* when he was eleven years old" (Sheffield *Borderlands* 15). However, as usual in what has always been a world movement, such inspirations and predictions are not limited to one man. To continue drawing examples from the exploration of space, I have already mentioned Wells and Serviss as inspiring Goddard. Jerry Pournelle (one of many SF writers who was also a scientist) recalls, "Most of my work was military aerospace, but I did get to work on Mercury, Gemini, and Apollo. We were helping to make the dream come true!" In Pournelle's case, the dream came, in large part, from Robert A. Heinlein. Pournelle says, "I devoured everything of his I could find, through high school, the army, college." Charles Sheffield, who was the only person ever to be president of both the Science Fiction and Fantasy Writers of America (SFWA) and the American Astronautical Society, said, speaking from his own broad experience, "Today, you go and talk to astronauts, lots of them—probably forty or fifty per cent—say their original interest in space began with science fiction, not with Jules Verne ... not with H. G. Wells, normally, but with people like Robert Heinlein and Arthur C. Clarke and Isaac Asimov" ("The Birth and Death of Science Fiction").

Harold Kaufman constructed the first **ion drive** in 1959, but Marc Rayman, chief mission engineer of Deep Space 1, which launched on 24 October 1998, gives an excellent example of the interconnectedness of science fact and science fiction, "I worked on a mission called Deep Space 1, which was the first interplanetary mission to use ion propulsion to travel around the solar system. And the first time I ever heard of ion propulsion was in the *Star Trek* episode 'Spock's Brain.' ... So the opportunity to connect what I saw in *Star Trek* as a little kid to what I'm doing now as an adult is very, very exciting" (Shatner).

To take a few examples from cybernetics and electronics, in 1898, in a story called "From the 'London Times' of 1904," Mark Twain suggested something that looked very much like a step in the direction of the internet, "The telelectroscope ... was soon connected with the telephonic systems of the whole world. The improved 'limitless-distance' telephone was presently introduced, and the daily doings of the globe made visible to everybody, and audibly discussible, too, by witnesses separated by any number of leagues." It's possible to argue that Twain had merely imagined the effects of a videophone. However, by 1946, Murray Leinster had gone considerably further. Murray Leinster (William F. Jenkins), who is well known to science fiction fans but relatively little known to everybody else, made truly extraordinary predictions in his short story "A Logic Named Joe." Leinster wrote about home computers, the net, and an integrated information, entertainment, and communication system. Although his computers are called logics and the net is called the tank, he describes a world where "ninety-four percent of all telecast programs ... all information on weather, plane schedules, special sales, employment opportunities and news ... all person-to-person contacts over wires ... every business conversation and agreement" (272), plus "all the facts in creation" (262) come from the tank by way of logics, which consist of "a vision receiver" with "keys instead of dials" (262). No one else had even begun to imagine the possibility of home computers linked to the internet at that early date, but amazingly, more than fifty years in advance, Leinster's story presented an accurate picture of what the net would become.

Sometimes, as with the world wide web, the time between prediction and creation is long, and the connection between the two events is uncertain, but on occasion the time is short and the connection obvious. Isaac Asimov's first collection of robot stories, *I, Robot*, was published in 1950, just four years after Leinster's story. In terms of scientific progress, the results were nearly immediate, "Joseph Engelberger ... built the first industrial robot, called Unimate, in 1958" crediting "his long-standing fascination with robots to his reading of *I, Robot* when he was a teenager" (Asimov, Warrick, Greenberg 69). A similar story involves the original *Star Trek* and Martin Cooper, inventor of the cell phone, though in this case, the time lag was more than twenty years. As Cooper tells it, "And suddenly, there's Captain Kirk talking on his communicator, talking with no dialing. That was not a fantasy to us, although to the rest of the world it was, but to me that was an objective" (Shatner). According to Shedroff and Noessel, "The connection was made even more apparent by the product's name: the StarTAC. The phone was a commercial success, arguably aided by the fact that audiences had been seeing it promoted in the

Introduction   19

form of *Star Trek* episodes and had been pretrained in its use for three decades" (6). Indeed, there is something to be said for the notion that SF pretrains us for life in the future.

I quoted Martin Cooper from a little film called *How William Shatner Changed the World*. It is partly humorous, playing on the supposed size of William Shatner's ego, and it allows him to be self-deprecating as he explains that the real message is How *Star Trek* Changed the World. More importantly, it signals an alteration not only in the speed of change (which had been increasing at an ever greater rate from the point when science and science fiction became major factors in the world), but also in the basic attitude to change. Asimov's starting point for science fiction when "the rate of change ... becomes great enough to be detected in ... an individual lifetime" (18) is a very different notion indeed from the idea that one person (or one television show which aired for three seasons) can transform the world and that that transformation can be, in a very real sense, deliberate. Nor is it a matter of a few writers and television shows. The size of the process is immense. J. M. Roberts says, "There are more scientists alive and working in the world today than have lived, worked, and died in the whole of human history hitherto." The speed and nature of change have become—except perhaps for a literature whose stock in trade is imagination—very nearly unimaginable.

Guy Lidbetter, Chief Technology Officer of Atos Global Managed Services, says, "Change is accelerating at an ever faster pace. From the advent of the colour TV to video recorders was about 15 years. The DVD followed in about 10. Then came Blueray HD in five and 3D TV in just three. We cannot expect this to slow." As Ray Kurzweil, inventor and futurist, explains it, "My models show that we are doubling the paradigm-shift rate every decade.... Thus the twentieth century was gradually speeding up to today's rate of progress; its achievements, therefore, were equivalent to about twenty years of progress at the rate in 2000. We'll make another twenty years of progress in just fourteen years (by 2014) and then do the same again in only seven years" (*Singularity* 11). Kurzweil suggests that progress in the twenty-first century will be "one thousand times greater than what was achieved in the twentieth" (*Singularity* 11), and even if his estimates prove unrealistic, it is still clear that the speed of change is enormous and that it is steadily accelerating in what may be, even for SF writers and futurists, unpredictable ways.

As I said in the Preface, from science fiction to fantasy, from magic to science and back again, humans have always been good at turning dreams to reality. One of the things we do best, as Ray Bradbury said in "The Toynbee Convector," is "To weave dreams and put brains and ideas and flesh and the truly real beneath the dreams" (14). And now we stand at a point where our new realities may transcend even our oldest, wildest imaginings. If we can "keep the utopian kernal ... safe in our imaginations with hope" (Zipes xiii), we may yet survive and prosper. The future we face has tremendous perils and transcendent opportunities. We will need our most powerful visions, our most rational thoughts, and our most numinous and magical words to see us safely through. This book is an exploration of those terms and perhaps a means of unlocking the door to the future that those terms are helping to create. (An earlier version of some parts of this appeared in my Introduction to *The Fantastic Made Visible*.)

Part One: The Terms of Science and Its Fictions

Alter Ego

The *Oxford English Dictionary* Online gives as its definition 2, "A person's secondary or alternative personality; a persona." And it adds, "Now the most common sense." How it got to be the most common sense is an interesting story, typical in some ways of the interactions of science and science fiction. The idea of alternative personalities or doubles is a very old one, and Shakespeare, who had already written two plays about twins, used it in *The Tempest* in 1611, a work that is generally considered to be **proto-science fiction**. The wizard Prospero is bracketed by Caliban and Ariel. Caliban (whose name is an anagram of cannibal in the carefree spelling of the English Renaissance) demonstrates the depths of violence and lust to which Prospero might fall. Ariel, a spirit of fire and air, shows Prospero the heights of compassion that he might reach.

While Shakespeare was, as usual, ahead of his time, the most frequently used Modern meaning effectively began on the edges of science, was explored in the wonder tales of E. T. A. Hoffman, and then cemented in the common imagination by Robert Louis Stevenson's science fiction novella *The Strange Case of Dr. Jekyll and Mr. Hyde*. David L. Pedersen says, "This 'modern' concept of duality starts in the 1730s when Anton Mesmer demonstrated the presence of 'another self' when he used ... 'hypnosis' to separate the alter ego.... Suddenly, people could see another character in an altered state of consciousness but in the same body. The theme exploded" (20).

Exploded indeed—it is not too much to say that there was an extended dialogue between the writers of **science fiction** and the theorists and practitioners of what would become psychology. Sharon Packer writes, "Nearly a century before Freud formulated his ideas about psychology, psychic doubles, and psychological projections, another writer made equally important observations that were destined to influence Freud's theories (and ballet, opera, theater, and film)." She means, of course, E. T. A. Hoffmann, and she goes on to say, "Freud acknowledged Hoffmann in his essay on *The Uncanny* and in *The Interpretation of Dreams*, and he admitted that poets and writers achieve insights into human behavior long before psychologists and scientists" (143). While Freud's praise of writers in general and Hoffmann in particular is well deserved, the process was more a dialogue than a source of inspiration that went one way. "We now know ... that Hoffmann's studies in psychic disorder ... profited from his familiarity with the latest in early nineteenth-century psychiatry" (Adler xi).

Many of Hoffmann's stories and novels deal with the idea of the double or alter ego. Perhaps the best known, because it became the inspiration for a Tchaikovsky ballet, is "Nutcracker and the King of Mice," written in 1816. But it was *The Devil's Elixirs*, first published a year earlier, that helped to set up a pattern that has now become either an archetype or a cliché, depending on the perspective of the critic, that of contrasting and warring selves, one good and one evil.

In the words of Patrick Labriola, "In *The Devil's Elixirs*, the double functions as an "id" who carries out devious actions." However, Hoffmann had anticipated Stevenson even more clearly. As Jeremy Adler puts it, "The dual character *Dr. Jekyll and Mr. Hyde* recalls Hoffmann's René Cardillac in *Mademoiselle de Scudery*, the demonic red-haired craftsman who at night exchanges his persona as the finest jeweler in Paris for that of a malignant murderer" (ix).

Robert Louis Stevenson's science fiction novella *The Strange Case of Dr. Jekyll and Mr. Hyde*, first published in 1886, tells the story of Jekyll's scientific experiment. "Through the ingestion of a concoction of powders, tincture, and salt," he releases his alter ego, "the thoroughly wicked Mr. Hyde" (Levin xii). The power of the story is generated by many elements, "But," as James Pope-Hennessy writes in his biography of Stevenson, "it was the idea of the total incarnation of a good man's evil nature into a Caliban figure which fascinated his Victorian readers, who were as preoccupied with good and evil as they were with death itself" (182). And it was not merely the physical incarnation of evil; it was also the struggle for survival, as Jekyll and Hyde strove with each other for control of their shared body.

The terror of the story for many readers was that Hyde became steadily more powerful while Jekyll grew ever weaker. Hyde's personality became the dominant one, and Jekyll felt himself being pushed out. "I became," he wrote, "...a creature eaten up and ... solely occupied by one thought: the horror of my other self" (90).

The dialogue between literature and psychology continued, but now there was a pop culture icon at the center of many of the conversations. In the English language alone, *The Strange Case of Dr. Jekyll and Mr. Hyde* was transformed into plays, musicals, radio dramas, television shows and more than thirty films, including comic versions by Jerry Lewis and Eddie Murphy, and a semi-feminist, semi-humorous version called *Dr. Jekyll and Ms. Hyde*. It seemed to illustrate or validate a variety of psychological theories. The *Great Soviet Encyclopedia*, for instance, called it "a classic working out of the theme of the 'split personality'" (vol. 24, 552). *Der Spiegel*, the German news magazine, used it when they wanted to explain Donald Trump, "There is no good Dr.

Jekyll behind the evil Mr. Hyde. Donald Trump is Hyde, the monster minus Jekyll" (Pitzke).

Though Stevenson's work was not inspired by Carl Jung's theory of the dark side of the self or shadow, the story almost seemed to be its ideal demonstration. In Jungian psychological theory, the alter ego is a remnant of the human evolutionary climb that must be integrated into the complete self. The 1953 film *The Neanderthal Man* used a similar approach but without any attempt at integration, the danger at the heart of almost all such stories. Professor Groves sets out to recreate a primitive man, saying, "Man has lost nothing of his emotions from the dawn of his history. He's lost nothing of his greed, his fury, his savagery, his jungle rapacity. Why not his physical personality?" Like Jekyll, he uses a chemical formula, and his primitive alter ego is, as usual, far stronger and more violent than he is himself. He reports, "All my basic animal instincts were enlarged and inflamed."

Jungian theory also argues that to deny the dark side risks an overwhelming of the self by the alter ego as is the case with Dr. Jekyll and his avatars. The film *Superman III* presents a mirror image of this problem when the evil Superman (who has denied any good impulses) suddenly finds his alter ego (good, of course) has taken physical form. After a prolonged struggle, the evil Superman is killed and replaced by the alter ego he had totally denied. This leaves the now entirely good Superman with a problem the filmmakers did not address—he has no alter ego and is, therefore, not a fully functional human being.

For example, in the Original *Star Trek* episode "The Enemy Within," where the transporter splits Captain Kirk into good and evil halves, the good Captain Kirk is unable to make decisions that would inflict pain, even when those decisions are necessary for the survival of his crew. His two halves must be reintegrated for the episode to end successfully.

Modern SF (and its pop culture analogues) has continued the practice of representing the alter ego of one character by another. In *Star Wars*, for instance, Luke Skywalker's alter ego is Darth Vader; Gollum is Frodo's alter ego, and the same may be readily said for Voldemort and Harry Potter. It might perhaps be possible to extend the metaphor to contemporary American politics as some commentators have done; Lowry, e.g., saw Newt Gingrich as Bill Clinton's shadow self. Jungian theory also sees the alter ego as a guide in the dark places of the psyche or elsewhere—on the second Death Star, in Mordor, or (to continue the political example) in the House of Representatives. A more lighthearted use of the alter ego in politics was bumper stickers that read "Republicans for Voldemort."

There are, also many examples of the struggle between the self and the alter ego in contemporary science fiction novels. Isaac Asimov's *The Robots of Dawn* pits two roboticists against each other, the good Han Fastolfe and the evil Kelden Amadiro. As one might expect, Amadiro seeks to take Fastolfe's place with his surrogate (**Robot**) son, his actual daughter, and his political constituency. Beyond that, the two of them are fighting over the usual small questions that arise in such stories—the settling of the galaxy and the fate of the human race. Again, as one might expect, the result is not a "crushing victory" for Fastolfe but a compromise which results in the two of them working together in Amadiro's "Robotics Institute" (380). After all, neither can survive without the other.

Recently, science has uncovered yet another (and even more primitive) source for the duality (or complexity) of human actions. Writing about "*individuality*" in 2015, Nigel Goldenfeld points out that a human is "a composite" and "in some senses not even human" since there are "perhaps 100 trillion bacterial cells" in a human body, "numbering 10 times more than ... human cells." More importantly, such bacteria "self-organize into communities" or "microbiomes." The "gastrointestinal microbiome can generate small molecules that may be able to pass through the blood-brain barrier." Such molecules "may be a significant factor in mental states such as depression" (27). Alanna Collen in *10% Human* (2015) provides some statistics, "Your gut alone hosts 100 trillion of them, like a coral reef growing on the rugged seabed that is your intestine. Around 4,000 different species carve out their own little niches.... Over your lifetime, you will play host to bugs the equivalent weight of five African elephants. Your skin is crawling with them. There are more on your fingertip than there are people in Britain" (2–3). Goldenfeld's conclusion makes the idea of a dual personality seem somehow familiar and safe, "You may be a collective property arising from the close interactions of your constituents" (26–27).

Anachronism

Means getting the time wrong, almost always, of necessity in realistic fiction and normal human lives, an earlier time, but with science fiction, it is also possible to get a future time wrong. Shakespeare, for example, in *Julius Caesar* puts pockets in togas, books with pages in those pockets, and clocks in Caesar's Rome. In *Antony and Cleopatra*, Shakespeare has Cleopatra playing billiards. But attitudes toward time and timelines were changing. Ben Jonson, perhaps the second greatest playwright of the English Renaissance, seems always to have been bothered by Shakespeare's casual attitude to sources and plots. Jonson's Roman play *Sejanus* was performed in 1603 with Shakespeare as a principal member of the cast and very possibly as coauthor. Robert Payne maintains that the "so happy genius" Jonson mentions in his introduction to the play "can hardly be anyone but Shakespeare" (260). However, by the time the play was published (and that introduction written) in 1605, Jonson had replaced all the lines of his collaborator with his own and provided a large number of Latin footnotes to justify his accuracy.

According to the *OED*, the first use of the word "anachronism" comes in 1646. By 1705, Thomas Hearne, antiquarian and historian, had criticized one of his society's most revered poets when he pointed out that "Virgil making Dido and Aeneas co-temporaries wheras they lived at three hundred years distance" (7) was an anachronism. Hearne defined the term as "a civil expression of an error or falsehood in chronology," distinguishing it from the "rude charge of falsehood" (7; spelling modernized).

Changing attitudes toward time were driving science and history and edging

Anachronism 25

toward the creation of science fiction. In part as a result, Hearne's attitude has now been almost completely reversed, and the kind of mistake that produces an anachronism is generally considered to be worse than fictional falsehoods committed deliberately for the sake of a particular story. Accurate knowledge has become a primary value, and the best defense against the charge of an error in chronology begins with the words, "Of course, I knew that...."

However, most anachronisms are obvious mistakes. The James Bond *technothriller GoldenEye* is a textbook example. It shows the complexities involved with even very small historical details and the difficulties in avoiding errors. Released for the Christmas season in 1995, the film dealt with the then new post–Soviet Russia. The opening credits show fallen and discarded statues, mostly of Lenin, but there are also huge representations of hammers and sickles. Hammers, sickles, red stars and one massive statue of Lenin fall. Soviet flags blow in the wind and then blow away. Finally, in one of the best symbolic depictions of the transition from Soviet Union to Russian Federation that I've seen, fashion models attack the statues with sledgehammers.

As always for Bond films (and books), the British secret service in general and James Bond in particular are presented as efficient and mostly well-informed. While watching one of several live satellite feeds, M says, "Unlike the American government, we prefer not to get our bad news from CNN." As usual with *technothrillers*, the audience is meant to believe that what they are seeing is accurate. This should apply especially to the Soviet and Russian information since these are spies, who, like Bond, have years of firsthand experience with the subject.

Enter General Ourumov, head of Space Division and one of the film's major characters. He appears first in the attack at Severnaya, and his uniform is obviously wrong. Hammers and sickles may have been tossed into a statues' graveyard at the beginning of the movie, but he still has them. The karakul fur hat he is wearing has a red-star emblem that contains a hammer and sickle. The buttons on his Soviet-issue winter coat also have stars with the hammer and sickle inside them. According to Colonel Girin, beginning in 1991, Russian uniforms were redesigned and the hammer and sickle eliminated. Winter coats, which hadn't been changed in twenty years, were made roomier, "which allowed them to be combined with a sweater, warm vest, and bullet-proof vest without the loss of mobility" (Girin). The old coats were sold on the international market as what has been called "Peace Surplus." They were inexpensively available to Eon Productions and to some members of the movie's first audiences (I had one).

When General Ourumov attends the council meeting in St. Petersburg, he is wearing a summer uniform, and again, his hat has the Soviet-era emblem, complete with a red star and hammer and sickle. There should be a golden star and above that a double-headed eagle, official symbol of the Russian Federation (Girin). Clearly some of the research on the film was slipshod, and the treatment of Russian Federation uniforms and emblems was careless to the point of insult. Worse still, the use of old Soviet uniforms in the first Bond film after the end of the Soviet Union (and the first one in six years) undercut the movie's message that Bond was not what M called him, just "a relic of the Cold War."

As with **technothrillers**, anachronisms in other kinds of science fiction can come in two categories since the advent of SF has made possible a whole new kind of anachronism—getting a future time wrong. Admittedly, such mistakes cannot be nailed on the wall for all to see with the same satisfying certainty as their past and gone cousins, but some of the elements in SF narratives clearly qualify. For example, in Paul Ford's 2015 story "The Last Museum," which is about the increasing speed of obsolescence, an intern describes climbing the 400 foot tall statue of Steve Jobs. Nearly everything has changed; nearly everything is connected but in different ways than most people had predicted. All physical stores are gone. Nevertheless, as he makes his escape after climbing up and then back down the statue, "the largest single continuous screen-clad surface in the world," the intern finds himself hiding in a laundry truck "with dirty towels from the staff gyms." And yes, it was always a good joke, from the time when Falstaff first hid in a laundry basket in *The Merry Wives of Windsor*, but with all the physical stores closed, how realistic is it that there would still be laundry trucks?

In the *Star Trek: The Next Generation* episode "The Measure of a Man," Data is referred to as a "toaster." It is unlikely to the point of impossibility that anyone (except a historian) whose food was produced by a replicator would know about such a primitive device. Much science fiction suffers from similar continuity (to use a film term) problems, in part because writers do not always work out all the implications of the worlds they have created and in part because writers (and presumably their readers and viewers) have difficulty abandoning the familiar furniture of their daily lives. (See **Nanotechnology**.)

There are, of course, many examples of such failures. The *Babylon 5* television movie *A Call to Arms* showed military personnel wearing parkas trimmed with fur and other humans with old-style eyeglasses (not the good, Augmented Reality kind which would have let them read all sorts of data across their lenses). Both are nearly impossible 250 years from now in a high-tech future. In the same way, *Star Trek: The Next Generation* blandly assumes that computer work stations will sit forever on desktops. *Babylon 5* and *STNG* indicate that hundreds of years in the future, poker, American football and baseball will still be popular. This last set of assumptions is an example of probable anachronisms.

While some games such as chess (invented before the sixth century) and tennis (played by Henry VIII and mentioned by Rabelais in *Gargantua and Pantagruel*) have been around for many centuries, many others have suffered the fate of whist, which went from great popularity in the eighteenth and nineteenth centuries to comparative neglect in the twentieth, when it was replaced by its descendent contract bridge.

There are, of course, instances in SF where deliberate anachronisms are used to achieve important goals. The repeated poker games in *Star Trek: The Next Generation* fall into that category, providing a familiar environment for the viewers and an arena of interactions for the characters. As Data says about the physicists' poker game he stages on the holodeck that includes Stephen Hawking, Isaac Newton, and Albert Einstein, "It has proved to be most illuminating" ("Descent Part 1"). And Data is correct, even though it's necessary to ignore the fact that Isaac Newton is speaking with a very modern upper class accent (Received Pronunciation) that did not exist during his lifetime (Romaine 16–22). I suppose it would be possible to blame that on a glitch in the holodeck.

Artificial Intelligence (also AI)

The *OED* defines it as "The capacity of computers or other machines to exhibit or simulate intelligent behaviour; the field of study concerned with this. Abbreviated *AI*." Their first example is from 1955, but the concept goes back as far as Homer's *Iliad*. (See **Proto–Science Fiction**.) In Hephaestus' workshop, "There were golden handmaids also who worked for him, and were like real young women, with sense and reason, voice also and strength, and all the learning of the immortals" (Homer 293).

The term is almost frustratingly general and may mean anything from the routine functioning of an expert system to a superintelligent and newly sentient being, though the latter is—so far—limited to the speculations of futurists and SF authors. An additional frustration, not only in defining terms but in developing technology is that the meaning of the word "intelligence" seems, like the horizon, to move ever further away as the explorer advances. Thus, playing chess clearly required intelligence, but once it had been accomplished by a program and the method explained, it just as clearly did not. The same was true for defeating the world's best chess player, for driving a car, and for writing sports stories for the Associated Press. In the words of Alex Armstrong, "Whenever you make something work, you know how it works and it no longer seems intelligent." There are, however, many other terms swirling around Artificial Intelligence, modifying it, expanding it, and sometimes even helping to explain it. Two older terms divide AI into broad categories. Weak AI is any system, no matter how simple or complex, that does work which would otherwise require a human, but does not seek to replicate a thinking being. Strong AI, on the other hand, means to do just that. To quote Alex Armstrong again, "At its most extreme strong AI aims to create an artificial consciousness" (Armstrong). Exactly what a consciousness would be and whether a machine consciousness would differ, of necessity, from our own, are vital questions.

Two more recent terms (and abbreviations) help to clarify the possible goals of Strong AI. Artificial General Intelligence (AGI) is, roughly speaking, the achievement of human-level intelligence. Of course, defining what would qualify as human-level intelligence or at least setting a series of tests or tasks that would indicate it had been reached is difficult. One suggestion is that an entity possessing AGI would be able to complete an undergraduate college degree. Another suggestion is that such an entity should be able to enter a strange house and make coffee. The best fictional representation of such minds is in the robot stories of Isaac Asimov. Artificial Super Intelligence (ASI) is mental power greater than that of humans, perhaps far greater.

Few ideas are more controversial at the moment than Artificial Super Intelligence or as it's usually denoted (somewhat confusingly) in the popular press, AI. In July of 2015, "Over 1,000 high-profile artificial intelligence experts and leading researchers ... signed an open letter warning of a 'military artificial intelligence arms race' and calling for a ban on 'offensive autonomous weapons.'" The letter was "presented at the International Joint Conference on Artificial Intelligence in Buenos Aires, Argentina," and its famous signatories included "Tesla's Elon Musk, Apple co-founder Steve Wozniak, Google DeepMind chief executive Demis Hassabis and professor Stephen Hawking" (Gibbs).

Seven months earlier, Hawking had delivered a much broader warning: "The development of full artificial intelligence could spell the end of the human race" (Cellan-Jones). Elon Musk had previously given "a $10 million grant to the Future of Life Institute ... an organization that says it is 'working to mitigate existential risks facing humanity,' the ones that could arise 'from the development of human-level artificial intelligence'" (Ford "Our Fear").

Even books on the subject (or their subtitles) suggest coming disaster. George Zarkadakis' book is called *In Our Own Image: Savior or Destroyer? The History and Future of Artificial Intelligence* (2015), and James Barrat sounds even more pessimistic with *Our Final Invention: Artificial Intelligence and the End of the Human Era* (2013).

Such warnings are meant to slow or even stop our progress toward Artificial Super Intelligence, Artificial General Intelligence, and perhaps even toward Weak AI or ever more complex expert systems. However, warnings or even prohibitions are unlikely to work. Apart from the obvious benefits to people in general that such progress would provide, it also offers additional power and wealth to the elites. In James Barrat's words, "The best corporate and government AGI projects will seek the competitive advantage of secrecy.... Few countries or corporations would surrender this advantage, even if AGI development were outlawed." He continues, "The technology required for AGI is ubiquitous and multipurpose, and getting smaller all the time. It's difficult if not impossible to police its development" (151–152).

Science fiction has provided us with a great many examples of how the development of intelligent machines can go wrong, but there are also positive examples—Asimov's robots, Iain M. Banks's Minds, and eventually, the Machine from the television series *Person of Interest* (2011–2016). In that five-year thought experiment, Harold Finch, who built an Artificial Super Intelligence, and Samantha Groves (Root), who wants to set it free, struggle to defeat or convince each other. In spite of the fact that he is the Machine's creator, Finch gives voice to the fiercest criticisms of ASI. He says, "We cannot understand these intelligences. The best we can hope for is to survive them" ("Prophets"). He says, "Our moral system will never be mirrored by theirs because of the very simple reason that they are not human" ("The Cold War"). He says, "Friendliness is something human beings are born with. AIs are only born with objectives" ("Prophets"). Root, who sometimes seems less human than an AI and thinks that humans are mostly **bad code**, declares on the subject of the Machine, "She loves us, Harold. She taught me to value life" ("The Cold War").

Perhaps because of his fears, Finch sets out to educate the machine and endow it with human values. His long-term struggle, coupled with Root's idiosyncratic sympathy, results in an ASI that regards Harold as her father and feels sorrow at the death of Root. The Machine says, "You can't conceive of my grief because you can't experience it the way I do" ("Synecdoche").

However, the primary conflict in *Person of Interest* is between two AIs, the Machine and Samaritan. Samaritan is the embodiment of Finch's worst nightmares, a nonhuman intelligence poised to manipulate and possibly destroy humanity. When the two Artificial Super Intelligences go to war, the Machine is fighting for the values Finch and Root have taught "her" and for the humanness "she" has indirectly experienced. David Brin, after rejecting the feasibility of renunciation, tight regulation and fierce internal programming as mechanisms for controlling AIs, recommends, "If you fear a super smart, Skynet level AI getting too clever for us and running out of control, then give it rivals who are just as smart but who have a vested interest in preventing any one AI entity from becoming a would-be God." In other words, we could have most of the benefits while minimizing the risks. Brin points out, "Alas, this is a possibility almost never portrayed in Hollywood sci fi—except on the brilliant show Person of Interest—wherein equally brilliant computers stymie each other and this competition winds up saving humanity. The answer is not fewer AI. It is to have more of them!" ("The One Thing"). At the end of *Person of Interest*'s thought experiment, it might be that even Harold Finch would agree with that.

Nick Bostrom, Director of the Future of Humanity Institute at Oxford University, writes in *Superintelligence: Paths, Dangers, and Strategies*, "Machine superintelligence would create a substantial existential risk. But it would reduce many other existential risks. Risks from nature ... would be virtually eliminated" (230). And opportunities for humanity might be enormously expanded. Of course, the issues, benefits and dangers involved with Artificial Intelligence are very complicated and at this stage impossible to resolve with any certainty. Many other entries provide detailed examinations of the issues and complications. **Deep Blue** is the most philosophical discussion of the rise of ASIs, but the following entries are also relevant to the interactions of humans and machine intelligences: **Asimov's Three Laws of Robotics, Bad Code, Frankenstein Complex, Kludge, Neural Lace, Positronic Brain, Robot, Robotics, Stepford Wives** and **Uncanny Valley.**

Asimov's Three Laws of Robotics

1. A robot may not injure a human being, or, through inaction, allow a human being to come to harm. 2. A robot must obey the orders given to it by human beings except where such orders would conflict with the First Law. 3. A robot must protect its own exis-

tence as long as such protection does not conflict with the First or Second Law. In the March 1942 *Astounding* in a story titled "Runaround," as Asimov just happens to remember, "on page 100, in the first column, about one-third of the way down" are Asimov's Three Laws of Robotics in print for the first time (*Robot Visions*, 7). There are slightly different versions of the creation of the Laws. In his autobiography *In Memory Yet Green*, Asimov says simply, "This story was the first one in which I explicitly stated all three Laws of Robotics" (317). However, in *War with the Robots*, coedited by Asimov, Patricia S. Warrick and Martin H. Greenberg, it was John W. Campbell (editor of *Astounding*) who first came up with them, though "Campbell maintained that the laws were in the stories and he simply pulled them out" (209). It is here too that Asimov adds a word to the language and a profession to the list of careers—**robotics**.

Note, however, that Asimov's laws are not as rigid as they at first appear but utilize (especially in the later books, though by no means limited to them) a kind of fuzzy logic or sliding scale of values so that an irrational or unimportant command from a human (Second Law) would not necessarily take precedence over the survival of a robot (Third Law).

Even more interesting conflicts come into play when robots begin weighing the value of human lives—many versus few, all versus many. Daneel Olivaw, Asimov's greatest robot creation, is doing something very like this as early as 1954 in *The Caves of Steel*. However, it is not until 1985 in *Robots and Empire* that Daneel puts the "Zeroth Law" into words: "A robot may not injure humanity or, through inaction, allow humanity to come to harm" (291). As he goes on to say, "The First Law should then be stated: 'A robot may not injure a human being or, through inaction, allow a human being to come to harm, unless this would violate the Zeroth Law of Robotics" (291).

Whatever the twenty-one year old Asimov may have had in mind beyond selling stories, his Three Laws changed science fiction and went on to change the world. Often enough, the two forms of change have coincided. Clearly, Asimov wanted to challenge what he called the **Frankenstein complex** and get rid of "the hordes of clanking, murderous robots" (*Robot Visions* 6). His position was that robots were machines and that "safety factors would be built in" (7). His robots were constructed by engineers. The problems they had were technological, not metaphysical. They had ceased to be symbols of human hubris and had become just one more sign of progress. As Asimov says, "The old-fashioned robot story was virtually killed in all science fiction stories above the comic-strip level" (8–9).

Good, safe robots became the norm. In Clifford Simak's *City* stories (originally published in *Astounding* between 1944 and 1947 and then combined into a novel) robots are the servants (and later saviors) of humans and dogs. In "Huddling Place," the second story in the *fixup*, robots help to make possible "the sort of life that men had yearned for years to have. A manorial existence, based on old family homes and leisurely acres, with **atomics** supplying power and robots in place of serfs" (*City* 36).

In the 1956 film *Forbidden Planet*,

Robby the Robot clearly obeys the three laws. Early in the film, Morbius, the Prospero figure in this space version of *The Tempest*, denies an implied accusation from Doctor Ostrow that relates directly to Robby. Morbius has indicated that Robby's strength is great enough that "he could quite easily topple this house off its foundation." Ostrow asks, "In the wrong hands, mightn't such a tool become a deadly weapon?" Morbius replies, "No, Doctor, not even though I were the mad scientist of the tape thrillers. Because, you see, there happens to be a built-in safety factor." It is at this point that Morbius demonstrates Robby's "absolute selfless obedience" and his inability to harm human beings. When Morbius orders Robby to fire a blaster at Commander Adams, Robby freezes, caught between the conflicting demands of the First and Second Laws. Only the cancellation of the order allows him to function again.

In the words of Chapman and Cull, *Forbidden Planet*'s "most significant borrowing from published SF was the robot Robby. Robby was every nut and bolt an extension of the fiction involving robots that was pioneered by Isaac Asimov ... as anthologized in the seminal 1950 collection *I, Robot*, the first story of which concerns a robot servant also called Robby" (82–83). Asimov spelled it Robbie, but *Forbidden Planet*'s powerful yet safe household servant (and by the end of the film, astrogator) was an excellent representative of the robot as useful machine. He (or it) later appeared in many movies and television shows, including *The Invisible Boy*, which was a kind of sequel to *Forbidden Planet*, and an episode of *Columbo* called "Mind Over Mayhem." It is a measure of Asimov's success that a seven-foot tall robot could appear in a murder mystery without for a moment being considered one of the suspects.

Isaac Asimov, who has written nonfiction as well as fiction about robots, became, in a sense, the world's most famous theoretical roboticist. The android Data from *Star Trek: The Next Generation* is, we are told in "Datalore," the result of Doctor Noonien Soong's desire "to make Asimov's dream of a **positronic brain** come true." In fiction at least, his Laws have even been used on people. In the "Phoenix Rising" episode of *Babylon Five*, Bester, a government agent and telepath, explains to Garibaldi the means he had used to control Garibaldi's actions, "I hit you with an Asimov. He was a writer long ago who wrote stories about robots. He came up with a set of rules to prevent them from turning against mankind.... Before we finished adjusting you, I made sure we planted Asimov's two most important rules in your mind. You cannot harm me directly or, through inaction, allow harm to come to me." Indeed, Asimov had already suggested that the Laws might be helpful in human behavior, but not in the way Bester uses them (Asimov, "Guest Commentary" 18).

However, at least as important as Asimov's success in changing the way robots were regarded in fiction was the effect his ideas had in the real world. "Joseph Engelberger, who built the first industrial robot, called Unimate, in 1958, attributes his long-standing fascination with robots to his reading of *I, Robot* when he was a teenager" (*War with the Robots* 69). Asimov says, "I have met other roboticists such as Marvin Minsky and Shimon Y. Nof, who also admitted, cheerfully, the value of their early reading of my robot stories" (*Robot Visions* 9).

With the coming of actual robots, however, Asimov's thought experiments in robot ethics took on greater importance. His stories had always pushed and pulled at the boundaries of the Three Laws. In the words of Rodney A. Brooks in his book *Flesh and Machines: How Robots Will Change Us*, "At first sight, the laws seem innocuous and plain common sense. However, upon closer examination they turn out to be very subtle. Of course, Asimov knew this, as he played on that subtlety for his plots" (73).

As Asimov clearly understood, laws mean what they are interpreted to mean. Robots are commanded to protect and obey humans, but as Asimov repeatedly demonstrates, the meaning of the word "human" can be problematic. In *The Robots of Dawn*, Vasilia orders four of her robots to dismantle the humaniform robot Daneel Olivaw. They cannot, however, move to obey her command until Daneel admits he is a robot and not the human being he resembles (293). It is not surprising that the robots are confused.

In *The Caves of Steel*, Lije Baley, the best detective in Asimov's various universes, repeatedly comes to the conclusion that Daneel is a human and not the robot he claims to be. Eventually, Baley gets to the truth, but it's a long process. In *Robots and Empire*, the Solarians redefine the term human for their robots so that it means, effectively, a human being who speaks "like a Solarian" (142).

There is a still more sinister redefinition of the term in Asimov's "... That Thou Art Mindful of Him," a story that he admits is an anomaly for him "since it is clearly a Robot-as-Menace story" (*The Complete Robot* 493). In it, the two robots George Nine and George Ten redefine themselves: "We find ourselves to be human beings within the meaning of the Three Laws, and human beings, moreover, to be given priority over those others" (*The Complete Robot* 517). Not only are "those others" actual humans, but the two rebellious robots plan to build "robots more advanced than ourselves" (517).

Asimov also frequently explored the troubling notion of the word "harm." In "Mirror Image," Lije Baley solves a mystery by manipulating two robots. Each robot is defending the reputation of its master, and Baley describes the possible harms differently for each robot. It becomes clear that in ordinary circumstances, a robot's behavior can be altered merely by altering its perception of possible damage to a human.

In *The Robots of Dawn*, Daneel and the roboticist Vasilia engage in an extended verbal battle over the ownership of the telepathic robot Giskard. Vasilia argues that she is being harmed by being deprived of Giskard, and Daneel argues that Giskard's current owner, Gladia, could not have him taken away without suffering even greater harm. Complicating the situation is the notion that Giskard should use his telepathic powers to make the woman who does not get to keep him forget about the whole matter.

Again, it becomes obvious that the concept of harm within the Three Laws remains unclear and is therefore largely subjective in the minds of robots. Daneel wins the argument against (or at least survives the onslaught of) a roboticist by appealing to an even more subjective concept—the good of humanity. This is the introduction of Asimov's Zeroth Law, and even Giskard, who is on Daneel's side in all this, points out the dangerous miasma of the idea, saying, "It is precisely because humanity is an abstraction that it can be called upon so freely to justify anything at all" (291).

Asimov had already demonstrated the hazardously subjective nature of the concept of harm in the Three Laws in a much earlier story about a telepathic robot—"Liar." In that story, Herbie lies to robot psychologist Susan Calvin. Herbie tells her that the man she secretly loves returns that feeling. He (or it) lies to other people as well. Dr. Calvin explains his motivation, "Do you suppose that if asked a question, it wouldn't give exactly the answer that one wants to hear? Wouldn't any other answer hurt us, and wouldn't Herbie know that?" (*The Complete Robot* 281). In revenge, Calvin destroys the robot by presenting him with a dilemma in which he must hurt a human being no matter what he does. She declares, "You can scrap him now—because he'll never speak again" (283).

Incidentally, if a robot that responds to people's emotions seems unrealistic, I point out that "Advertising giant M&C Saatchi is currently testing advertising billboards with hidden Microsoft Kinect cameras that read viewers' emotions and react according to whether a person's facial expression is happy, sad or neutral" (McStay).

Asimov's writing about robots—fiction and nonfiction—represents more than fifty years of thought experiments, a long, amazingly clever set of extrapolations and interrogations. His Three (or Four) Laws are often the starting point even for people who disagree with him. AI theorist Ben Goertzel, for example, says, "The point of the Three Laws was to fail in interesting ways; that's what made most of the stories involving them interesting." He continues, "So the Three Laws were instructive in terms of teaching us how any attempt to legislate ethics in terms of specific rules is bound to fall apart and have various loopholes" (Dvorsky).

Hans Morevac, on the other hand, in discussing the robot-controlled industries he foresees, argues that "We voters should mandate installation of an elaborate analog of Isaac Asimov's 'Laws of Robotics' in the corporate character of every powerful intelligent machine" (140).

Indeed, Asimov asked a set of tough ethical and scientific questions. Here are some of the more important ones: 1. What does it take to be a good person, robot or human? (*The Robots of Dawn*) 2. If there are two sentient species on Earth, does one have the right to own and control the other? ("The Bicentennial Man") 3. Can deontology (a law-based system of ethics) work effectively? (*Robots and Empire*) 4. Will robots eventually transcend and even supersede their creators? ("... That Thou Art Mindful of Him") 5. Will robots and humans become ever more alike until they are, in fact, a hybrid species? ("Segregationist").

What happens next in the real world of robots will be part of the development of **artificial intelligence** and even perhaps of the coming of the Singularity, Vernor Vinge's term for the point in time when we create intelligences greater than our own. James Barrat writes in *Our Final Invention: Artificial Intelligence and the End of the Human Era*, "If we're going to try and imbue an AI with friendliness or any moral quality or safeguard, we need to know how it works at a high resolution level before it is able to modify itself. Once that starts, our input may be irrelevant" (66). Asimov's Laws provide at least a beginning for that knowledge.

Atomics

The *OED* defines it as, "The science or study of atoms, esp. in relation to atomic energy and atomic weapons." In *Brave New Words: The Oxford Dictionary of Science Fiction*, Jeff Prucher divides it into two words, one meaning "any device powered by nuclear fission" (11) and the other meaning "atomic weapons" (12). In *Futurespeak*, Roberta Rogow gets closer to the way the word works in SF, "Anything that involves the use of nuclear fuel or energy as a power source" (21).

Note that the switch from atom to nuclear tends to be a reflection of a shift from studying an atom as an entity to studying its internal parts or nucleus. However, if we start with "the science or study of atoms" (which in the nineteenth century had meant chemistry) or even nuclear physics, it is all too easy to forget that the idea behind atomics was power, power to destroy everything or power to create a utopia with all the trappings that people thought of as modern or imagined as futuristic. It was to be the atomic age. It was and still is both worst nightmare and perfect dream.

Science fiction writers and futurists, those people who looked to tomorrow with longing, who imagined transformations that most people would not accept even as fictions, began thinking about the idea behind atomics very early. In 1870, Jules Verne's Captain Nemo, master of what was clearly a ship and a vision of the future, described the power that made his extraordinary submarine possible, "There is a powerful agent, obedient, rapid, facile, which can be put to any use and reigns supreme on board my ship. It does everything. It illuminates our ship, it warms us, it is the soul of our mechanical apparatus. This agent is—electricity" (Miller and Walter 77–78). At the time, Verne's claim was impossible—and unexplained. But in the future he imagined, there would be a power source to generate enough electricity to do everything. Eventually (1954), there would be a real *Nautilus* named, in part, after Verne's fictional one, the world's first operational nuclear submarine.

In 1895 in his novel *The Crack of Doom*, Robert Cromie described the power that was to be found in matter, "One grain of matter contains sufficient energy, if etherised, to raise a hundred thousand tons nearly two miles. In face of such potentiality it is not wise to wreck incautiously even the atoms of a molecule." At the end of chapter 11, Herbert Brande, the mad (or at least seriously unbalanced) scientist who had delivered that warning, sets off the fictional world's first atomic blast: "The sea behind us burst into a flame, followed by the sound of an explosion so frightful that we were almost stunned by it. A huge mass of water, torn up in a solid block, was hurled into the air, and there it broke into a hundred roaring cataracts. These ... fell ... into the seething cauldron of water that raged below." And all this "was really caused by a drop of water" torn apart for the energy it contained.

In 1909, Garritt P. Serviss, an SF writer who was far less famous than Jules Verne, made a prediction as prescient as some of the latter's. Serviss was an ordinary writer with extraordinary ideas. (See the Introduction.) He imagined "atomic energy liberated and harnessed to drive a rocket to the planet Venus" (Searles). In the words of A. Langley Searles, "His conception is uncannily close to truth; he names uranium as the raw material from which is extracted the vital substance, a 'crystallized powder' which releases its energy on proper treatment." In *A Columbus of Space*, Serviss writes, "Every one of those minute grains ... is packed with as much potential energy as that of a ton's weight suspended a mile above the earth." Of course, Serviss's prognostications were not without foundation. In 1904, to take one example, Ernest Rutherford had written in his groundbreaking *Radio-Activity*, "If it were ever found possible to control at will the rate of disintegration of the radio-elements, an enormous amount of energy could be obtained from a small quantity of matter" (338).

In 1914, H. G. Wells began his novel *The World Set Free* by saying, "The history of mankind is the history of the attainment of external power" (*The Last War* 1). Wells' story was both a prediction and a warning, though the warning had a sting of hope in its tail. It was also one of the clearest examples of the reciprocity between science and science fiction. Wells based his work on that of "such scientific men as Ramsay, Rutherford, and Soddy" (18). Indeed, the book is dedicated to "Frederick Soddy's 'Interpretation of radium'" since Wells' novel "owes long passages to the eleventh chapter of that book." Frederick Soddy, who won the 1921 Nobel Prize in Chemistry for his work in radioactivity and isotopes (which he named in 1913) went on to write in 1926 (in *Wealth, Virtual Wealth, and Debt*) about energy and how it created real wealth by turning materials into physical goods. In that book, Soddy maintained that if there was "an unlimited source of energy of the order of a million times more powerful than any known, what tremendous social consequences await the discovery of artificial transmutation!" He continued, "In the *The World Set Free* Mr. H. G. Wells ... devoted himself with his customary brilliance and insight to the question, and so vividly depicted the probable consequences that it would be superfluous for anyone of lesser gifts to pursue the topic, at least until the practical realisation of the disturbing dream comes nearer" (28). Wells' warning was of atomic bombs, and his bit of hope was that the world would (having used them in one war) give up such monstrous futilities and make a world government instead.

Wells' novel presents "atomic engineering" as a path to plenty, a largesse beyond the capacity of humanity to understand or enjoy, "In the full tide of an incalculable abundance, when everything necessary to satisfy human needs and everything necessary to realise such will and purpose as existed then in human hearts was already at hand, one has still to tell of hardship, famine, anger, confusion, conflict, and incoherent suffering" (27). As so often, Wells is at least as concerned about the mechanisms of society as he is about the inventions of science, "There was no scheme for the distribution of this vast new wealth that had come at last within the reach of men; there was no clear conception that any such distribution was possible" (27). What would come instead (by 1956 Wells suggested) was war, but he had joined himself to those who looked forward (however illogically) to a new world of abundance driven by atomic engines. And ironically, he might well have helped to shorten the time it took to get atomic power and bombs and thus have invalidated the timeline of his own prediction. He was hoping for (in some areas at least) a self-defeating prophecy. Instead he got not only a self-fulfilling prophecy but a self-accelerating one.

It was 1933. Leo Szilard, the brilliant Hungarian who had trained in physics in Berlin and shared patents with Einstein, was then a refugee in London. His intellectual furniture included Wells' *The World Set Free* with its bombs that created not one but a series of explosions. He was angry because he had read a newspaper story about a physics conference in which Ernest Rutherford had said what came to be a sort of litany for him that there was no way to get large quantities of energy from the atom. It was "moonshine" (Jenkin 128). Szilard said, "As the light changed to green and I crossed the street ... it suddenly occurred to me that if we could find an element which is split by neutrons and which would emit *two* neutrons when it absorbs *one* neutron, such an element ... could sustain a nuclear chain reaction" (cited in Rhodes). The result could be (in fact, as it turned out, would be) "to set up a nuclear chain reaction, liberate energy on an industrial scale, and construct atomic bombs" (cited in Rhodes). Szilard said, "This possibility intrigued me so much that I gave up the idea of shifting to biology" (cited in Lanouette). In the words of William Lanouette, "He credits this decision to three people: H. G. Wells, 'who showed Szilard 'what the liberation of atomic energy on a large scale would mean'; and Frederic and Irene Joliot Curie ... who ... demonstrated that radioactivity could be created artificially."

H. G. Wells had warned against the dangers of radioactivity (and perhaps suggested chain reactions) even before he wrote *The World Set Free*. In 1908, in *Tono-Bungay*, Wells has Gordon-Nasmyth describe "quap" as "the most radio-active stuff in the world." He says, "There it lies in two heaps, one small, one great, and the world for miles about it is blasted and scorched and dead" (227). Later in the novel, George Ponderevo, the narrator, raises this to a species of nightmare, "To my mind radio-activity is a real disease of matter. Moreover, it is a contagious disease.... You bring those debased and crumbling atoms near others and those too presently catch the trick of swinging themselves out of coherent existence" (336). At last he declares, "I am haunted by a grotesque fancy of the ultimate eating away and dry-rotting and dispersal of all our world" (337).

There were other writers who sounded warnings. By 1937, J. B. Priestley had grasped the sheer size of the change that was coming. Of George Glenway Hooker, one of the main characters in his novel *The Doomsday Men*, Priestley wrote, "He spent his working hours in a magical world of infinitely minute solar systems, where the very elements themselves can be transmuted. He and his friends had forced their way into the secret laboratory of Nature herself. It was quite possible ... that one of them might emerge, with a handful of mathematical and chemical symbols, to announce that the universe that houses us is quite different from what we have so far imagined, and might prove what many have guessed at, namely, that Newton, with his solid engine of a cosmos, was wrong, and that Shakespeare, with his dissolving towers and palaces and universal stuff of dream, was strangely right."

What Priestley concocted was, as his title indicates, an atomic **doomsday machine**. Most of the mechanisms (including the particles) are, of necessity, imaginary, but in the novel, physicist Paul MacMichael discusses how he will destroy the world by bombarding a new heavy element which he has discovered and named paulium after himself with what he calls dynatrons, another of his discoveries. He declares, "I think I can promise instant dissolution." In the end, the world is saved, but Priestley has described something that sounds uncomfortably like a cross between Wells' "contagious disease" of matter and a cobalt bomb.

And there was, of course, much more. "As early as 1942, Lester del Rey wrote a novella (published in *Astounding*) about a meltdown at a nuclear power plant." Called "Nerves," it "presupposes a world where atomic power is an accepted fact, where atomic-power plants are as commonplace as automobile factories" (Healey and McComas 47). That was "three years before the first bomb and twelve years before the world's first 'nuclear powered electricity generator began operation' in Obninsk, Russia" (Pilkington "Forbidden Planet" 51).

In 1946, Pat Frank (who, in 1959, would write *Alas, Babylon*, a much more serious—and far better known example of what are called *fictions of nuclear disaster*) produced a comic novel titled *Mr. Adam*. In it (and the comic part comes later), "The great new nuclear fission plants at Bohrville, Mississippi ... disintegrated in an explosion that made Nagasaki and Hiroshima mere cap pistols by comparison" (13). Poul Anderson's 1947 story "Tomorrow's Children," suggested that a nuclear war would mean worldwide mutations for plants, animals, and humans, so many in fact that in all probability the mutants "will outnumber the humans" (124). The story also predicted a nuclear winter. "The last three winters had come early and stayed long. Dust, colloidal dust of the bombs, suspended in the atmosphere and cutting down the solar constant by a deadly percent or two ... *Fimbulwinter*" (117–118).

Even for hypothetical uses in space, there was concern about atomic power. In *Across the Space Frontier*, edited by Cornelius Ryan (which started with the First Annual Symposium on Space Travel held at the Hayden Planetarium in New York in 1951), no one was recommending the nuclear option. In "Prelude to Space Travel," Wernher von Braun discussed several problems with atomic power, including "the weight of an atomic pile" and "radioactive contamination of the launching site and the necessity of radioactive shielding." He does say, though, "The outlook for atomic rocket propulsion appears somewhat brighter for interplanetary voyages originating and terminating at a space station in an orbit above the earth" (22).

For Willy Ley, however, in "A Station in Space," the station itself would be better off with solar than with atomic power, "Even a small atomic pile has an enormous weight, and what it finally produces, in addition to radioactive substances for which the station would have very little use, is again just steam or vapor" (112). In fact, if Cornelius Ryan's Introduction is to be taken at face value, his 1952 book was "an urgent warning that the United States should immediately embark on a long-range development to secure for the West 'space superiority' since a ruthless power established on a space station could subjugate the world" (xiii). The subjugation would come by means of "projectiles armed with atomic war heads," so though he never suggests that an American Space Station would be anything other than an observation platform, it is just possible that in Ryan's hypothetical future, there might have been a nuclear option after all.

The discoveries about and the debate over atomics did not stop in the 1950s, and they continue even now. The pendulum has swung from positive to negative and back with what sometimes looks more like wild abandon than sober decision making. In February of 1950, Leo Szilard warned "that it would be 'very easy to rig an H-bomb' to produce 'very dangerous radioactivity.' All you had to do said Szilard, was surround the bomb with a chemical element such as cobalt that absorbs radiation. When it exploded, the bomb would spew radioactive dust into the air like an artificial volcano.... 'Everyone would be killed,' he said" (P.D. Smith xvii–xvii). Here was Priestley's science fiction presented as science fact. In 1960, Herman Kahn was indicating it as one more scenario on a list of options (Ghamari-Tabrizi 211–212).

One of the stranger forays into nuclear planning was Project Plowshare. Part of the "Atoms for Peace" movement, which had everything from support from then–President Eisenhower to "an hour-long television program called Our Friend the Atom" produced by Walt Disney (Kaufman 20), Plowshare resulted in a number of experiments (twenty-seven nuclear explosions by the end of the program) and truly grandiose plans, including the "Pan-Atomic Canal," which was to be blasted through Nicaragua at sea level. An earlier thought experiment had considered replacing the Suez Canal with a new waterway "all the way across Israel, from the Gulf of Aqaba to the Mediterranean" (Regis "What Could").

When the Soviets created the world's first ICBM and launched Sputnik in 1957, Project Plowshare's planners suggested a demonstration of American geo-engineering power as a reminder that the Russians weren't the only ones who could do amazing things. "'Why not,' Lewis L. Strauss, chairman of the Atomic Energy Commission, was asked, 'find a reasonably remote place (such as Alaska, perhaps) which needs a harbor, and do the job with one blast?' Furthermore it made sense to have both the media and the public witness it as proof of American technological skill" (Kaufman 24). Named Project Chariot, the Alaska harbor demonstration came closer to realization than most of Plowshare's ideas. The experiment was shelved, in part, because of objections from Native Americans and the Russians. Both groups believed that they would be harmed by the radiation generated by the "six bombs" (Regis *Monsters* xiv) that would make up that one blast. Elon Musk's suggestion six decades later that one way to terraform Mars is with atom bombs appears to be in the tradition of Plowshare. He told Stephen Colbert in 2015, "The fast way is drop thermonuclear weapons over the poles" (O'Callaghan "Elon Musk").

Of all the uses to which the sheer power of atomics might be put, perhaps the most interesting for futurists and SF writers was the exploration of space. In spite of the objections to the radiation that would be generated and the rejection of the option by many space scientists, there were still those who went ahead with plans for atomic spaceships.

The most extraordinary idea was for "Project Orion ... an interplanetary space ship powered by nuclear bombs" (Dyson 2). "Suggested missions ranged from the ability to deliver 'a hydrogen warhead so large that it would devastate a country...' to a grand tour of the solar system" (Dyson 5–6). Expectations were enormous, and many people thought that the plans would be realized. However, "Project Orion's dreams of large-scale transport of passengers and freight around the solar system" depended on the development of "small, clean, fission free or extremely low fission bombs" (Dyson 11). And this did not happen. The main legacy of Orion was the atomic spaceships that blasted off in science fiction stories and films, like the *Botany Bay* in the Original *Star Trek* episode "Space Seed" or the fake but supposedly plausible atomic vessel in the television series *Ascension*.

Beyond the many predictions of disaster, accidents with real nuclear power plants at Three-Mile Island, Chernobyl and Fukushima, plus the fear of radiation in general, have dimmed the luster of atomic power from what Kimberly Wells (Jane Fonda's character in the nuclear disaster film *The China Syndrome*) describes as "that almost magical transformation of matter into energy." For instance, in March of 2011, "Over 40,000 protesters in southern Germany formed a 45-kilometer (28-mile) human chain from the Neckarwestheim nuclear plant to the nearby city of Stuttgart, demonstrating against plans that extended the life of Germany's nuclear power stations" (Grässler, Levitz, and Knight). Sigmar Gabriel, leader of Germany's Social Democrats, said, putting his party's spin on the situation, "We are experiencing an end to the atomic era" (Zuvela).

The issue is hotly contested. It would be hard to find a better expression of the other side than that of Senator Arnold Vinick in the television series *The West Wing*, "You know why Europe's Greenhouse emissions are so much lower than ours? Nuclear power, totally emissions free. You know how many Americans die from oil refinery explosions, from coal soot in the air? Tens of thousands. And not one from a nuclear power anything in thirty years."

In one of the most extreme condemnations of atomics ever written, Ed Regis, in a 2015 *Slate* article and the book behind it, labels both Project Plowshare and Elon Musk's suggestion for terraforming Mars "pathological technology." He sees it as "what amounted to an overriding, all-consuming, and almost irresistible emotional infatuation" (*Monsters* xiii). But while that may be true of his first example, the *Hindenburg* and other zeppelins filled with "millions of cubic feet of explosive gas" (twenty-six of which had been destroyed in similar fashion before the *Hindenburg* disaster), it is a far too facile dismissal of the hopes, studies and struggles behind nuclear power.

For example, the Soviet city of "Pripyat was built in 1970 as an ideal socialist city for the 50,000 workers employed at the Chernobyl nuclear plant. It was considered a privilege to live close to the reactors." There is still "a propaganda slogan praising the civilian application of nuclear energy ... displayed high up on a crumbling concrete-block building on Lenin Square." It says, sadly for those who read it now, "The atom is a worker, not a soldier!" (Schmundt and Thoma).

Nevertheless, the effects of the Chernobyl disaster have been far less deadly than many people expected. Unit 4 exploded on April 26, 1986. "Firefighters tried to extinguish the flames and to cover the open reactor core. Many ... were exposed to extremely high doses of radiation and, by 1998, 39 of them had died" (Dworschak). The biggest problem since then has been thyroid cancer among children, with 6,000 cases, 15 of them fatal (Dworschak). Beyond that, "Higher cancer rates in the population have thus far not been determined." At least that was the conclusion of "the United Nations Scientific Committee on the Effects of Atomic Radiation (UNSCEAR) in 2011" (Dworschak).

Meanwhile, "Chernobyl today" appears to be "a nature paradise." Not surprisingly, "the still-elevated radiation seems to be less damaging to nature than humans are." This does not, of course, take into account how severe the effects were and continue to be. Timothy Mousseau, a biologist who has studied the Chernobyl area for years, says, "Literally every rock we turn over, we find a signal of the mutagenic properties of the radiation in the region" (Zimmermann).

One argument that emerged from the tragedy of Fukushima was whether or not most people (small children excepted) would have been safer in their own homes (or in the hospitals or hospices where they were), despite the radiation, than in the massive (nearly 100,000 people) evacuation. "A study conducted by the University of Stanford concluded that there were 600 victims of the evacuation, compared to the maybe 30 that would have died of radiation poisoning had they not been rescued" (Dworschak).

I am not arguing that radiation is safe or that nuclear power plant meltdowns are minor problems. I live in an area that is "downwind" from the Nevada test site, and my father-in-law, who is a chemistry professor, was one of the experts who was on call for and might have been sent to Chernobyl. I am merely suggesting that nuclear fission has both dangers and advantages and that it should now be evaluated not as dream or nightmare but as a fact that has emerged from (though it continues to be shaped by) the imaginations of scientists and the fictions of writers.

Some people are even now suggesting that a new generation of nuclear power plants, smaller, more efficient and far safer, may bring us closer to that utopia of atomics. Taylor Wilson, among others, has talked about modular power plants that are "built in a factory," run on "highly enriched uranium and weapons-grade plutonium that's been down-blended," and "buried below ground." They would "run for 30 years without refueling," and they "operate at essentially atmospheric pressure, so there's no inclination for the fission products to leave the reactor in the event of an accident."

Still, almost everyone (Taylor Wilson included) who still sees atomics as the future is looking forward to nuclear fusion. And for perhaps the first time, nuclear fusion is looking like it really will happen: It was a scene that would have fit beautifully into a science fiction film. A woman who is the head of a powerful country and who is also a physicist stands ready to throw the switch on an experiment that if it is successful promises unlimited energy with minimal costs and few environmental problems. The experiment was the result of a 1.1 billion-dollar investment that took nineteen years to prepare. And, as we would expect in a film, "The success achieved exceeded even the most optimistic expectations" ("New Issue"). It was February 3, 2016, the woman throwing the switch was Dr. Angela Merkel, and in the experiment, the Max Planck Institute's Wendelstein 7-X (W7X) stellerator "managed to sustain a hydrogen plasma," a very big step forward (Andrews). One commentator on YouTube wrote, "Help us Wendelstein 7-X, you're our only hope" ("Fusion Reactor").

Of course, there are many other ventures, most spectacularly, the ITER Project in the South of France, jointly funded by the European Union, China, India, Japan, Russia, South Korea, and the United States. The plan for ITER is that it will, for the first time in history, "achieve a fusion reaction that produces more energy than it consumes ... at least 500 megawatts of power from a 50 megawatt input" (Clery).

There will not be fusion generators any time soon, but it's safe to say that they are coming, and however strange and magical they may seem, they will have little to do with Jedi masters, but will be the result of long thought and much imagination, of the work of futurists and SF writers, and of a costly research effort that spans decades and occupies laboratories across the world. At last, the power of atomics, dreamed of, hoped for and many times promised, will be delivered.

Bad Code

A flawed design, originally having to do with computers but used by the television series *Person of Interest* to mean humans as well. As applied to humans, it seems also to reflect (though there is no quotation of this in the series) the IBM acronym BAD "'Broken As Designed' ... Said of a program that is bogus because of bad design and misfeatures rather than because of bugginess" ("BAD"). (See **Kludge**.) The term is part of a philosophical disagreement between the character Harold Finch, the moral center of the series and creator of an **artificial intelligence**, and Samantha Groves (or Root, as she calls herself), who is, at least initially, his amoral **alter ego**. It concerns the nature of human beings, the future (if any) of human evolution, and the role AIs should play now and in times to come. The debate is carried on in the midst of kidnappings, torture and murders, many of them committed by Root.

In "The Contingency" (2012), Root compares humans with Finch's AI, "One day I realized—all the dumb and selfish things people do—it's not our fault. No one designed us. We're just an accident, Harold. We're just bad code. But the thing you built—it's perfect, rational, beautiful by design." Root expands on her view of humanity in the "Bad Code" episode, "We still haven't upgraded human beings. The human race is stalled out, Harold, and from what I've seen, most of it is rotten to the core." Finch disagrees with her, pointing to John Reese, the show's hero. "You're wrong. He proves you're wrong. Not all humans are bad code." But Root has plenty of evidence on her side. Denton Weeks, one of the people in the government who has access to Finch's AI, tries to kill him. Root saves Finch's life with the comment, "What did I say, Harold? Bad code" ("Bad Code").

In fact, even Finch has trouble being optimistic about humans all the time. In the episode "Triggerman," he talks to John Reese about the mob enforcer Riley Cavanaugh, saying, "Why are you wasting time on Riley? He's a killer. He's just bad code." However, Riley gives his life to save the woman he loves, and when, near the end of the episode, Reese asks Finch what he had meant by 'bad code,' Finch backtracks from what had sounded very much like Root's position, "It means a flawed design. The term applies to machines, not to people. We have the ability to change, evolve." (See **Mutant**.)

That is, on one level at least, the central argument between Root and Finch and the meaning of bad code, at least as Root (and through her series) defines it. She says, "The real reason you built the machine is because the world is boring. Human beings have come as far as we're gonna go. I want to see what happens next" ("The Contingency").

At this point, Samantha Groves sounds very much like Isaac Asimov, who wrote in his Introduction to *War with the Robots*, "We might cynically suggest ... that humanity deserves to be superseded; that its record as custodian of the Earth has been a miserable one ... so that it is time another form of life was tried. We might reason that it is a great honor that humanity has proved worthy of designing its own superior successor. We might even maintain that our great fear should be that this successor will not be produced soon enough to save the Earth, and that every effort should be bent to replacing our miserable species as soon as possible" (6–7). Or in the words of Dolores, an android "Host" from the 2016 HBO series *Westworld*, "This world doesn't belong to you, or the people who came before. It belongs to someone who is yet to come." (See **Robot**.)

BEM (Bug-Eyed Monster)

A humorous acronym (and term) first used in 1939, BEMs were old fashioned space aliens with no redeeming social value who looked more or less like insects and existed to threaten the hero and make the heroine scream. They were also meant to appear on the covers of magazines and books to help sales, and more importantly, they provided enemies who could be killed in large numbers with no concern for the violence involved and no trace of guilt.

Robert A. Heinlein's *Starship Troopers* (1959, filmed in 1997) with its war against "the Bugs" is an example, as are *Predator* (1987) and its sequels, where the aliens have mandibles, and also the "Dalek" storylines of *Doctor Who*.

In the 2005 *Doctor Who* episode "The Parting of the Ways," Rose Tyler, the Doctor's companion (or her "Bad Wolf" **alter ego**), completely destroys the Dalek Emperor and the new generation of Daleks, even though the new Daleks are made from human cells. She says, "I can see the whole of time and space, every single atom of your existence, and I divide them. Everything must come to dust." No one, not even the often obnoxiously moral Doctor, questions their elimination.

More enlightened stories began as early as 1949 with Fredric Brown's "All Good Bems." They featured misunderstood bug-eyed monsters such as the Horta in the Original *Star Trek* episode "The Devil in the Dark" or at least tough, nonscreaming heroines as in the *Aliens* series. Heinlein's 1958 novel *Have Space Suit Will Travel* includes a sympathetic character who is called "The Mother Thing" (242) but was first described as "a bug-eyed monster" (43). Kip (the book's narrator) immediately says, "That's not fair but it was my first thought" (43).

In the **first contact** novel *The Mote in God's Eye*, Larry Niven and Jerry Pournelle use the term to explore the attitudes of people to the first space aliens the human race has ever met. One of the aliens says, "We play your part in order to understand you, but you each seem to play a thousand parts. It makes things difficult for an honest, hard-working bug-eyed monster." One of the humans, concerned that their intolerant history should be so clearly exposed, asks, "Who told you about bug-eyed monsters?" And the answer is, "Mr. Renner, who else? I took it as a compliment—that he would trust my sense of humor, that is" (234).

Indeed, it is a joke for the readers of the novel as well as for the aliens. Science fiction fans will recognize the history of the phrase, which includes a certain amount of ridicule directed at unsophisticated depictions of nonhumans, with parody anthologies such as *Bug-Eyed Monsters and Bimbos* (Resnick). But Niven and Pournelle also give us Commander John Cargill, who uses the term in all seriousness, "Captain, I don't know. I don't know a lot of things about those bug-eyed monsters" (280). And they give us Horace Bury, whose analysis of how to deal with the Moties is, "*Get them before they get us!*" (399). As Niven and Pournelle make clear, the history of the idea of bug-eyed monsters is also, on some level, the history of how humans think about aliens. Fortunately, those thoughts seem to be becoming ever more complex and, with some obvious exceptions, humane. (See **First Contact**.)

Big Dumb Objects

Are gigantic, nearly inexplicable alien artifacts which have (usually) been abandoned by their makers. These may be planetary systems for transforming thoughts into matter like the one the Krell create in the film *Forbidden Planet*, enormous space ships as in Arthur C. Clarke's *Rendezvous with Rama* and David Brin's *Startide Rising*, a wide range of weird and dangerous things as in Charles Sheffield's Heritage Universe series and Gregory Benford's Galactic Center series, an abandoned vacation home as in Jack McDevitt's *Ancient Shores*, or even a Dyson Sphere as in Bob Shaw's *Orbitsville* and *STNG*'s "Relics."

Often left behind by superpowerful aliens such as Forerunners or Ancients, they provide technologies like the Stargates (in *SG-1* and *Stargate Atlantis*) that are far beyond the capabilities of humans. Obviously, there are dangers when Big Dumb Objects are operated by small dumb people. In some cases, the BDOs are abandoned because they caused the extinction of their makers, though there is hardly ever a warning to that effect. Even the Krell, great geniuses with a million years of civilization behind them, were destroyed by their own system, which created whatever their conscious or subconscious minds imagined, with no filter or control whatsoever—a Big Dumb Object indeed.

The term was first used by Roz Kaveny in 1981, and then, as an April Fool's joke in 1992, by Peter Nicholls, who was writing an entry for *The Encyclopedia of Science Fiction*. He says, "I decided that I would write alien artifacts but call it big dumb objects, and write in a poker-faced style, suggesting an even more absurd critical term to be used in its place, 'megalotropic sf'" (Nicholls "Big Dumb Objects"). However, even in print BDOs are dangerous. Nicholls continues, "The joke was on me, because as I came to write the entry, I realized that the subject—which was vast alien enigmatic artefacts–was at the heart of what attracted people to science fiction." It's a long article, actually delivered as a lecture, and Nicholls ends with something that points the way to the place where Big Dumb Objects really came from. He maintains that "writers of hard science fiction are as a group probably more not less romantic than their soft-sf colleagues; that this romance is to a degree intrinsic in the very metaphors that deep space produces, ranging from space itself through enigmatic alien artefacts to the furthest reaches of cosmological speculation." (See **Hard SF**.)

It is then the romance of science and exploration, of knowledge so vast that it takes the breath away (especially if your space suit happens to fail at the wrong moment). It goes back to the true beginning of science and science fiction, to the time of the Enlightenment and the French Revolution.

In 1798, Napoleon Bonaparte invaded Egypt. He brought with him, in addition to his army, "167 geographers, botanists, chemists, antiquaries, engineers, historians, printers, astronomers, zoologists, painters, musicians, sculptors, architects, Orientalists, mathematicians, economists, journalists, civil engineers, and balloonists" (Andrew Roberts 165). The inclusion of this secondary army of scholars was not only an expression of the Revolution's desire to learn everything, controlling the world with knowledge, but also of Napoleon's urge to imitate the expedition of his hero Alexander the Great.

In all of human history, there is no better example of encounters with Big Dumb Objects. The sheer power of the ancient Egyptian civilization, whose artifacts were all around them but whose words and purposes were lost in seemingly immeasurable time, stunned everyone. "Napoleon often said thereafter that 'of all the objects that had impressed him in his life, the pyramids of Egypt ... were those that had most astonished him'" (172). It is one of those indelible historical experiences that echoes, like the fall of the Roman Empire, far into the future and the dreams of the future.

By its very nature, such an experience is about the long vistas of time, a natural habitat of science fiction. So, the end of Rome echoes in tales of Galactic Empires, and the Great Pyramid of Giza, which was the oldest of the Seven Wonders of the World and the tallest structure on Earth until the twentieth century, is transmuted to something truly alien and transplanted to a far distant star system. In the words of Shelley, "Round the decay/ Of that colossal wreck, boundless and bare/ The lone and level sands stretch far away" (546).

Bobble

In Vernor Vinge's *Peace War* series, a stable "confinement sphere" or spherical force field which can enclose anything for a long period of time. The term force field seems to have been invented by John W. Campbell in *Islands of Space* (1931). In the book version of that novel, he wrote, "Arcot was heading for the magnetic force field which surrounded the city when Torlos made a mistake." Like Campbell's force field that can surround a city, bobbles are closed systems, with all of the possibilities that suggests. In *The Peace War* (1984), Vinge describes "what it's like inside a ... bobble": "Nothing comes in or goes out. If you explode a nuke in such a place, there is nowhere to cool off.... The innocent-seeming bobble buried in Tucson all these decades, contained the heart of a fireball."

As the series continues, the ability to manipulate the technology of bobbles advances to the point where a bobble can be given a termination date or a time when it will open, no matter how far distant in the future that might be. Bobbles are first used as weapons in various ways, but they have other possibilities as well. Since they are, in fact, stasis fields, not only sealed off from the outside but unchanging inside, they can also be used as what Robert A. Heinlein called "a sort of half-baked, horse-and-buggy time travel" (*The Door into Summer* 445). In other words, they can be used in the same way as **cryonics** to move forward in time without all the fuss of inventing a **time machine**.

In *Marooned in Realtime* (1986), Vinge says, "Inside the bobble, time was stopped. Those within were as they'd been at that instant of a near-forgotten war when the losers decided to escape to the future. No force could affect a bobble's contents; no force could affect its duration—not the heart of a star, not the heart of a lover" (7). Clearly, Vernor Vinge has produced one of the more interesting speculations on the nature and possibilities of force fields, far beyond the usual terse report that follows an attack, "Captain, the forward shields have been reduced by sixty percent," as though a barrier made entirely from energy will function in roughly the same way as armor plating.

Chronotransference

In Robert J. Sawyer's 1993 short story "Just Like Old Times," the term means to "project a human being's consciousness back in time, superimposing his or her mind over that of someone who lived in the past" (16). The technique has limited value since the consciousness cannot be retrieved, and the person from the future dies when the past individual he or she is inhabiting dies. In addition, the future consciousness is no more than a passenger, unable to influence the thoughts or behavior of the target mind.

In Sawyer's story, the drawbacks of the technology are turned into advantages. It becomes a popular method for euthanasia and in one case, the central narrative of the tale, an attempt by a judge at inflicting the death penalty. Unfortunately (but not unexpectedly for readers of time travel stories), the serial killer who is sent back into the limited consciousness of a T. rex learns "to control the beast," lengthening its life and putting him in position to hunt the small primates that were the ancestors of humans and therefore to hunt "every single human being who would ever exist" (25).

Sawyer uses the term again in "You See But You Do Not Observe" (1995), "I had been pulled into the future first, ahead of my companion. There was no sensation associated with the chronotransference, except for a popping of my ears." This time it seems to mean no more than a standard form of time travel, with some unspecified mechanism picking up the traveler and moving him, whole and unharmed, from the past to the future.

Pat Forde's term "deep-projection" in his short story "In Spirit" is actually closer to Sawyer's first use of chronotransference. Forde writes, "Deep-projection was not 'time travel' as anyone had foreseen it. It was more like stepping into a virtual-reality display of the past, a past that we cannot affect in any way.... At least, we believed that at the start. And for the most part it's perfectly true: we cannot affect the past in any way that can change what's already happened." Both chronotransference and deep projection seem to operate by sending information without a physical incarnation into the past, a far more plausible form of time travel than the usual variety.

The full name for the process Forde describes is "Transdimensional Extended Projection technology," but it's usually called "ghosting" because it amounts to sending a ghostly projection into the past, a nearly disembodied consciousness which can observe the past and move about in it but not influence it except by offering an odd sort of emotional comfort that has a positive effect on the people in the past even though it seems to come from nowhere and nobody. Forde suggests that the experience is also positive for the ghosts because it has resulted in "a wider recognition of our present as the precious, precarious climax to all our ancient pasts. After millennia of struggle and strife, back-breaking labor and bad luck, madness and sadness and small successes piled one atop the next, most societies are making that final leap up the ladder of progress toward a transcendent, tolerant civilization." I suspect Forde (and a number of other writers) might be willing to say something similar about the effect of time travel narratives and other historical fictions. (See *Speculative History*.)

Clarke's Laws

Arthur C. Clarke, like many science fiction writers, was also a futurist. His three laws are actually laws of scientific prediction, or as he put it in the Introduction to his nonfiction book *Profiles of the Future* (1963), "This book ... does not try to describe *the* future, but to define the boundaries within which possible futures must lie" (xi). His laws (in "Hazards of Prophecy: The Failure of Imagination") are part of that defining of boundaries, a part that evolved over the years in various editions of the book.

The first law was there from the beginning and was clearly identified as a law: 1. "When a distinguished but elderly scientist states that something is possible, he is almost certainly right. When he states that something is impossible, he is very probably wrong" (14).

What became the second law was also in the first edition of the book, but it didn't get promoted to the status of law until a later revision: 2. "The only way of discovering the limits of the possible is to venture a little way past them into the impossible" (20).

The third law was also present after a fashion, but again, it didn't become a law until later. Originally, Clarke had written, "No matter how far sighted and imaginative he might be, your pre-twentieth century scientist would have said: 'What utter nonsense! That's magic, not science'" (19).

In a footnote at the end of "The Hazards of Prophecy" chapter in the 1984 edition, Clarke identifies his Second Law with the comment, "Originally, I had only one ... but my French editor started numbering them." He continues, "The Third, which arises from the material in this chapter, is perhaps the most interesting and most quoted, 'Any sufficiently advanced technology is indistinguishable from magic'" (36). In *Futurespeak*, Roberta Rogow expands the third law with a useful phrase, "Any technology sufficiently advanced in relation to its observers is indistinguishable from magic" (52).

Ultimately, Clarke's laws are about the process of prediction, and as is only to be expected with such a rational observer, they are explorations (with copious examples) of how prophecies have gone wrong in the past and will, therefore, most likely go wrong again in the future.

Corpsicle

A person or other creature frozen and placed in cryosleep in order to be revived at some point in the future. The term was invented by Frederick Pohl in 1969 in his novel *The Age of the Pussyfoot*, "It is true, however, that no corpsicle has yet been thawed and returned to life" (210). The word is clearly a variant of popsicle (and icicle) and shows a certain level of disrespect for **cryonics** or at least for those who have participated in the process. Though Pohl's overall discussion of cryonics in the Author's Note to his novel is not negative, he is well aware that the frozen dead are at the mercy of the living.

Will McIntosh's short fiction from the January 2009 *Asimov's* introduces yet another variant—the story's title is "Bridesicle." Here again, as so often happens in fictions about the cryofuture, the bridesicles are at a distinct disadvantage. As one man explains to a prospective bride who has been only partially revived, "Your insurance covered the deep-freeze preservation, but full revival, especially when it involves extensive injury, is terribly costly. That's where the dating service comes in." In fact, for those women who do not have sufficient wealth to pay for their own resuscitations, marriage is the only way back to life. As another man says about marriage ceremonies in the cryonics facility, "They happen all the time here. It's kind of risky to revive someone otherwise." Obviously, these men are not much like the lovesick prince who could not live without seeing Snow-White in her glass coffin.

While McIntosh has come up with an interesting new term and yet another way for corpsicles to be exploited, I'm not sure from the story's background information why there are bridesicles but not groomsicles. Are we to assume that all cryonically frozen males have enough money to deal with their own revivals? In a society of scarcity, which this one clearly is, that's hardly possible. Are we then to assume that women would not have the money to pay for revivals if they found suitable husbands? Again, that seems highly unlikely. Perhaps we just didn't get to see the groomsicle section of the facility (or of another facility somewhere else) because the main character (and victim) was female. Or perhaps, as sometimes happens, an extrapolation of what will probably happen in the future hasn't completely thrown off the notions of what has happened so often in the past. In all fairness, though, McIntosh's story is more complex than my brief recounting of it. The main character, for instance, describes herself as a lesbian, and there are other major plot developments as well.

Cosmogony of the Future

Invented by Donald A. Wollheim in his 1971 book *The Universe Makers*, it is a consensus history of the future which many **science fiction** writers and futurists share, comprising the human exploration and settlement of the solar system and then the galaxy. Wollheim argued that "there is only a limited number of general possibilities open to human conjecture." He continued, "When all the many highly inventive minds of science-fiction writers find themselves falling again and again into similar patterns, we must perforce say that this does seem to be what all our mental computers state as the shape of the future" (42). Wollheim broke his future history down into eight steps and acknowledged that Isaac Asimov's Foundation series was "the point of departure." The beginning steps are indeed universally accepted and not only by SF writers, "First, we have the initial voyages to the moon and to the planets of our solar system.... Second, the first flights to the stars" (42).

The idea of moving outward from the Earth to the moon, then to the planets and asteroids, and finally to the stars is an old one, going back, in some form, to **proto-science fiction**. However, in the late nineteenth and twentieth centuries, it was arguably the future that almost everyone, from futurist to scientist to science fiction writer, expected or at least hoped for. It might at first seem that this was, more than anything else, a case of shared ideas proliferating, but two of the most important people to come to these conclusions—Tsiolkovsky and Goddard—did so in isolation.

Konstantin Tsiolkovsky, partly deaf and largely self-taught, had read Jules Verne, but his thoughts and his proofs were definitely his own. As William J. Walter says in *Space Age*, Tsiolkovsky "believed that someday more advanced versions of his reaction machine would scatter the human race across the Milky Way, and make humans the inhabiters of countless worlds. In the process of this great migration, he predicted, we would build self-sustaining space stations, explore the moon and Mars, and design earth-orbiting solar power stations, space suits, and artificial gravity machines. He imagined a time when humans would re-engineer the environments of other planets, populate the asteroid belt beyond Mars, and communicate with extraterrestrial beings" (7).

On January 12, 1920, the reclusive rocket scientist Robert H. Goddard found himself temporarily famous as the result of a casual suggestion he had included in a pamphlet of his (published by the Smithsonian) that a rocket might be able to reach the moon. He was ridiculed by newspapers across the country, but he had not included his more interesting "fantasies about 'operators' flying rockets controlled by gyroscope and side jets, communication with extraterrestrial beings, manned journeys to Mars, ion propulsion (he even had a patent on this)" to go with his patent for a liquid-fuel rocket (Walter 29).

By 1952, in his essay "Prelude to Space Travel," Wernher von Braun proposed a space station orbiting the Earth at a distance of 1,075 miles. From there, he argued, voyages to the moon and planets would not be impossible, "Voyages to the nearest planets such as Venus and Mars are entirely in the realm of possibility with presently known propellants, provided such trips begin and end in an orbit such as that 1,075 miles above the earth. Indeed the propellant requirements per pound of payload for a Mars-bound spaceship ... would not be much greater than the amount of fuels needed to leave the orbit for a trip to the much closer moon" (49).

The idea of space travel had moved from science fiction or isolated scientific eccentricity to a common dream, the cosmogony of the future, or at least its first steps.

Of course, in between then and now we entered the space age, put humans on the moon, and sent spacecraft throughout the solar system and beyond. Our success has been built on the work of Tsiolkovsky, Goddard, von Braun and tens of thousands of others, including many who were not engineers but were pioneers of the imagination, showing the way for the engineers who followed as Jules Verne had done.

Today, the future history that depicts our expansion through the solar system and then to the stars is part of our culture and indeed, is already happening. As I write this, Boeing is running a television commercial describing what things will be like in 2116 and beyond, titled "You Just Wait." It says, "Whole communities are living on Mars, and solar satellites provide Earth with unlimited clean power. In less than a century, Boeing took the world from seaplanes to space planes, across the universe and beyond. And if you thought that was amazing, you just wait" ("Boeing TV Spot").

While "the first flights to the stars" are not yet on anyone's drawing board, NASA and other groups are gathering the data that will eventually make such journeys plausible as well as possible. "In early May [2016], a team of astronomers led by MIT announced that they had discovered three Earth-like planets orbiting a small dwarf star, known as TRAPPIST-1 (named for the Transiting Planets and Planetesimals Small Telescope), only 40 light years away. The planets are considered the most likely worlds to find life outside our solar system due to their size, moderate temperatures, and their short distance from us which makes them easy to observe." Additional observations by the same team using the Hubble Space Telescope indicated "that the two innermost planets, TRAPPIST-1b and c, were rocky, but also that they have compact atmospheres similar to what you might find on Earth, Venus, or Mars" (Bennett "Pair of Planets").

This is only the latest in an extremely energetic search for exoplanets, a process that will become significantly more efficient in 2018 when NASA launches its next big telescope. The James Webb Space Telescope has a mirror (made up of eighteen smaller mirrors) "that is about three stories tall.... The base of the craft is as long as a tennis court" (Wenz). There are also multiple instruments that can see in the infrared. In the words of Jeff Coughlin, a NASA researcher, "I'd say the James Webb Space Telescope's main impact will be the ability to study the atmospheres of exoplanets in detail, down to sizes just a bit bigger than the Earth" (cited in Wenz).

The other six steps in Wollheim's list are more problematic and less widely accepted. They are: "Third, the Rise of the Galactic Empire.... Fourth, the Galactic Empire in full bloom, regardless of what form it takes.... Fifth, the Decline and Fall of the Galactic Empire.... Sixth, the Interregnum.... Seventh, the Rise of a Permanent Galactic Civilization.... Eighth, the Challenge to God" (43–44). The "Interregnum" represents a slide into societies so primitive as to lose (in many instances) the knowledge of spaceflight. And during the "Challenge to God" stage "Galactic harmony and an undreamed-of high level of knowledge leads to experiments in creation, to harmony between galactic clusters, and possible exploration of the other dimensions of existence" (44).

The assumptions behind the last six steps are harder to justify than those behind the first two, though Wollheim's explanations (which I have mostly omitted) are more open-ended than his initial list. For example, the galactic civilizations he suggests are impossible without some form of FTL (faster than light travel). However, he notes "the problem of whether science can ever exceed the speed of light—a very important one where the problem of colonization is concerned" (42).

In fact, the galactic empires he suggests seem much more like the empires that have existed on Earth than the ones that might be possible in space, and as a rule, they seem to struggle with the kinds of problems that afflict societies with limited resources and comparatively primitive technologies. For instance, step three includes, he says, "The rise of contact and commerce between many human-colonized worlds or many worlds of alien intelligences that have come to trust and do business with one another" (143).

Imagine a galaxy where almost all spacefaring species have **nanotechnology** and can produce literally anything they need with minimal effort from their own planets and solar systems. This does not preclude the possibilities of trade. In all probability, there will still be art and artifacts, patterns and technologies, and even interactions and friendships to be exchanged. But "commerce" as it currently exists on Earth is unlikely. There will be no container ships traveling between the stars, loaded with bulk cargoes of basic goods. Not only will nanotech produce anything that sentient creatures might need, it will also produce it anywhere they happen to be. Commerce as it currently exists will be largely unnecessary.

Just as the physical embodiments of books and music have largely been replaced by digital files, a library of digital instructions for making almost everything and assemblers to construct them when necessary will be the birthright of everyone who lives in an advanced civilization.

The problems of describing the histories of future societies have not discouraged futurists and SF writers, and many of Wollheim's eight steps have been (by one author or another) followed faithfully, expanded cleverly, or inverted and distorted ferociously. SF writers have included all sorts of technologies, some of them nearly unimaginable even to the people who were in the process of imagining them. While many of the works that have followed where Asimov and Wollheim went before might be described as space operas, they have, as space operas have always tended to do, been voyages of discovery and thought experiments illuminating future pathways and making, as Shakespeare said long ago, "wonder seem familiar," ultimately turning the future into a history that it seems—somehow—we have always known.

Cryonics

The freezing of humans and other creatures in order to facilitate their travel from one place or one time period to another, or to suspend their life processes until appropriate medical treatment is available to correct illness or injury, or until mortality may be upgraded to immortality. There is a sometimes heated debate about whether or not cryonic technology has now or will soon cross over from fiction to fact. There are at least a few hundred human bodies (and some heads) which have been cryogenically frozen and are being stored with the stated goal of unfreezing and reviving them when the technology (presumably **nanotechnology**) has advanced sufficiently.

The feasibility of that goal is what heats up the debate. Cryobiologists, scientists who study the effects of cold (from chilly to far below frozen) on various life forms, have, on occasion, spoken out against cryonics as a pseudoscience. This should not, of course, be confused with the successful use of extreme temperatures to preserve human embryos, ovarian tissue and other human cells.

According to Stephanie Watson, here is how those people (and heads) were frozen: once a person has been declared legally dead, a team arrives quickly, provides an anticoagulant for the person's blood, and supplies the "brain with enough oxygen and blood to preserve minimal function" until the client reaches the "suspension facility." The idea is that legal death need not mean the complete cessation of brain activity and therefore revival will be possible. Water, which would freeze and destroy cells, is replaced "with a glycerol-based chemical mixture called a cryoprotectant—a sort of human antifreeze." The bottom line is to get to storage in a tank filled with liquid nitrogen by stages. The most important part of "this process, called vitrification (deep cooling without freezing), puts the cells into a state of suspended animation."

This is radically different from what Robert A. Heinlein described in *The Door into Summer* (1956), "About four in the afternoon, with Pete's [a cat] flat head resting on my chest, I went happily to sleep again" (494). Apart from sharing space with the cat (which is not common even in the future Heinlein describes), this amounts to suspended animation (freezing) while the subject is alive (no draining the blood to replace it with antifreeze) and no being submerged in liquid nitrogen to maintain the temperature. Perhaps this is merely fantasy or perhaps our current methods are just very primitive. So far, of course, there have been no revivals with those current methods, and the procedure (primitive or otherwise) remains untested.

Cryonic suspension, whether of the recently deceased or of the still living, belongs to a broader category of SF story devices and thought experiments whose purpose is to "freeze" a character for a journey in time or space. As folklorist Andrew Lang said in an essay on the subject, "The early fancy of the ballad-mongers and fairy tale-tellers ... has dwelt longingly on the idea of suspended animation." Heinlein calls it "a sort of half-baked, horse-and-buggy time travel" (*The Door into Summer* 445). (See **Bobble**.) The simplest of these methods is sleep—extended, as in Washington Irving's "The Legend of Sleepy Hollow," for an unnaturally long period of time.

Other writers who found this method workable included L. S. Mercier in *L'An 2240* (1771, translated by W. Hooper in 1772 as *Memoirs of the Year Two Thousand Five Hundred* [sic]; see the Introduction), Edward Bellamy in *Looking Backward* (1888), and H. G. Wells in *When the Sleeper Wakes* (1899). Obviously, this has more to do with fiction than prediction, and a somewhat similar experience (in **fantasy** rather than SF) is that of humans who wander into the world of faery in folk tales and find that time there moves at a different rate of speed so that a few hours can amount to many years.

J. R. R. Tolkien's *The Lord of the Rings* borrows the notion of an "elven" world which is somehow outside time in Lothlorien, where there is no moon, that celestial marker of the passing months. In a way, any kind of immortality (or extremely long life) is also a form of time travel or freezing in place as the years spin by, much like what the Time Traveller sees in the 1960 film version of Wells' *The Time Machine* and (with less interesting special effects) in the original novel.

This is true for Flint in the *Star Trek* episode "Requiem for Methuselah," Professor John Oldman in Jerome Bixby's (also the author of the script for "Requiem") *The Man from Earth* (film and play), Enoch Wallace in Simak's *Way Station*, Simak's "Grotto of the Dancing Deer," the Ioan Gruffudd television series *Forever*, and in the *Highlander* films and television series.

Another form of freezing in place makes the experience entirely literal.

This is true in Mary Griffith's *Three Hundred Years Hence* (1836) and in W. Clark Russell's *The Frozen Pirate* (1887). The main character in *Three Hundred Years Hence* (the first utopian novel written by an American woman) was "buried in the fall of the ice and hill," but he thaws out and revives with no ill effects. The same is true for the title character in *The Frozen Pirate*. The narrator quite rightly finds the process unbelievable, "There was no power in fire to stretch him to his full length and turn him over on his back. What living or ghostly hand had done this thing? Did spirits walk this schooner after all?"

In fact, it is no more than a plot device that will be often repeated, even in times when almost everyone knows the destruction that water wreaks when it freezes inside a cell. For instance (to pick two interesting examples from a long list), the situation comedy *The Second Hundred Years* (1967–1968) features a grandfather frozen in a glacier and successfully brought back by nothing more sophisticated than warmth (at which point he discovers that he is "younger" than his son and the same "age" as his grandson). The same is essentially true of *Iceman* (1984), though in that case there is some talk of the frozen caveman's body somehow having produced a chemical which protected his cells.

More interesting than literal freezing as a means of preservation, is the idea of a chemical that suspends animation without ill effects. Grant Allen's "Pausodyne," (1881) and T. E. Pemberton's dramatic farce *Freezing a Mother-in-Law* both use the idea, though for very different purposes. In Grant Allen's story the "compound" is called Pausodyne, and "it possesses the singular power of suspending animation in men or animals for several hours." To his sorrow, the inventor discovers that Pausodyne can suspend animation for far longer than that. There is also an antidote, but working alone has its drawbacks. Pemberton's farce features chemicals with similar effects, "A few drops of this colorless liquid ... injected into lobe of ear of patient will freeze, or suspend animation of patient.... Remedy, few drops of another liquid injected into lobe of other ear, and warm bath." If long-term human storage proves as simple as this, Pemberton and Heinlein (among others) should get recognition for the prediction.

The *OED* credits an advertisement in *Galaxy Magazine* in 1966 with the first use of the word cryonics: "Immortality is within your grasp. The Cryonics Society of New York Inc. is the leading non-profit organization in the field of cryogenic interment." However, as this entry makes clear, the idea was around much earlier, as were several eminently usable alternative terms. Jeff Prucher points out in *Brave New Words* (24) that Robert A. Heinlein was the first to use cold sleep as a noun in the 1941 magazine publication of *Methuselah's Children*. The book publication (1958) changes the quotation substantially, "They put up with it only long enough to rig for cold-sleep" (111). In his 1956 novel *The Door into Summer*, Heinlein also uses cold sleep as a verb for the first time, "He was quite capable of refusing to let me cold-sleep" (350). In fact, he racks up quite a list of such terms, starting with cold sleep as a noun, "Until shortly before then, I could not have paid for cold sleep; it's expensive" (340).

Then, soon after, he has "Stasis, cold sleep, hibernation, hypothermia, reduced metabolism" (352). He also uses "Long Sleep" (345).

The *OED* says that the second use of cryonics is in Clifford Simak's 1977 novel *Heritage of Stars*, but this is wrong. The quote in both the original hardback and the current Kindle is, "You read about cryogenics: freezing the passengers and then reviving them" (183). In any event, the idea behind all these terms is the same, though the details vary with the imaginations and prognostications of the writers. In *Tomorrow and Tomorrow* (1997), Charles Sheffield seems to have coined revivables (20) and cryowombs (27). He also uses cryocorpse, but that term appears earlier in Lois McMaster Bujold's 1989 novel *Brothers in Arms* as cryo-corpse (10). Even more interesting is Sheffield's suggestion of using liquid helium instead of liquid nitrogen. He describes a cryonics facility that can "go to five thousand atmospheres pressure" and push water ice through various phases, "Then by the right combination of temperatures and pressures work all the way down toward absolute zero, passing into and through phases 5, 3, and 2" (22).

The cold plunge into reality or the possible reality of plunging into the cold was largely the result of R. C. W. Ettinger's efforts. "Persons cooled quickly enough and kept at low enough temperatures could be restored to life, said Ettinger in his 1964 Book-of-the-Month Club bestseller, *The Prospect of Immortality*" (Pascal). He wrote, "We need only arrange to have our bodies, *after we die*, stored in suitable freezers against the time when science may be able to help us. No matter what kills us, whether old age or disease, and even if freezing techniques are still crude when we die, *sooner or later* our friends of the future should be equal to the task of reviving and curing us. This is the essence of the main argument" (Ettinger 12). The book was published by Doubleday. They "sent a review copy to.... Isaac Asimov, who gave it a clean scientific bill of health. The book appeared in nine languages, four editions, and became the bible of the cryonics movement. Ettinger found himself appearing on *Time* and *Newsweek* and nationwide TV" (Pascal).

The reaction from science fiction writers was typical—they descended gleefully upon the sociological issues and practical difficulties of what had suddenly moved from interesting idea to possible reality. In *Why Call Them Back from Heaven?* (1967) by Clifford D. Simak, all of society has been taken over by a massive corporation whose stated goal is to freeze everyone until the day when a utopia featuring eternal life for all would be available. "There came the day when it became apparent that the little movement of 1964, now called Forever Center, had become the biggest thing the world had ever known" (47).

One of the questions that SF writers immediately pounced on was what would happen to the money and other holdings that belonged to the people who were frozen. Heinlein had already played with it in various ways in 1956 in *The Door into Summer*. In *Why Call Them Back from Heaven?*, "all those millions who now lay frozen had left their funds in trust with Forever Center. And one day the world woke to find that Forever Center was the largest stockholder of the world and ... it had gained control of vast industrial complexes" (47). It was beyond the regulation or control of any government. Ultimately, Simak presents it as an enormous scam that can never be debunked or overthrown. The book's last line is, "God has turned his back on us" (191).

Larry Niven's 1973 novella *The Defenseless Dead* suggests that frozen humans would be a good source from which to harvest spare parts or, as the story calls it, "organlegging."

Some visions of the cryonic future were not quite so bleak or at least not quite so sweeping. In *The Age of the Pussyfoot*, Frederick Pohl presents a world that revives the frozen dead but at very high cost, "Forester gasped and coughed and cried, half strangled, 'Two and a half million dollars for medical.... Who can afford that kind of money?' The response is, not surprisingly, "'Why you can, for one. Otherwise you'd still be frozen'" (65).

In *Tomorrow and Tomorrow*, Charles Sheffield suggests an even less sympathetic cryo-future, "Drake was convinced that even when a foolproof way of resuscitating the Revivables was discovered, most bodies in the cryowombs would stay exactly where they were." Drake, who wants his wife to be revived and cured of a fatal disease and wants to be revived himself so that he can continue his life with her, has to find a way around some difficult questions, "Why thaw anyone at all? Why add another person to a crowded world unless he or she had something special to offer?" (27). Why indeed? After all, what is the worth of a **corpsicle**?

Frederick Pohl said about cryonics in an Author's Note to *The Age of the Pussyfoot*, "Demonstrate that it works one time, and we'll grab for it as we've grabbed for few things before" (210). Grant that the technology can be perfected in the reasonably near future, and we will have a world with millions of people taking the Long Sleep. Will that mean some version of the problems SF writers and futurists have predicted? Lois McMaster Bujold's 2010 novel *Cryoburn* (along with the other books I've mentioned) suggests that it might.

There are even a few possibilities I haven't mentioned. In the Sylvester Stallone film *Demolition Man* (1993), John Spartan (Stallone) is "sentenced to 70 years ... sub-zero re-habilitation in the California Cryo-Penitentiary." This may not seem like much of a punishment or a means of rehabilitation either, but while he is here his "behavior will be altered through synaptic suggestion." A more likely possibility is indicated in *Iceman*, "Some day paramedics are going to carry cryotanks." If millions of people do commit their bodies to cryowombs, it will be a gesture of hope, a belief very much like Ettinger's that the future will be wiser, richer, and kinder than the present, that those desperate and desperately hopeful people from the past will be welcomed, their illnesses cured, and their lives extended as they pass through what Heinlein called the door into summer.

Cyberspace

A virtual reality version of the information highway entered by humans who directly interface (brain to chip, more or less) with computers, now used to refer to the infobahn itself and more mundane connections. The transition can be tracked in reference works. In

1993, the *Random House Unabridged Dictionary* defined cyberspace with three words, "See virtual reality" (497). By 2004, *The Penguin Encyclopedia* used only one, "Internet" (415). The term was first used by William Gibson in the 1982 short story "Burning Chrome" and the novel *Neuromancer* in 1984. The *OED* defines it as, "The space of virtual reality; the notional environment within which electronic communication (esp. via the Internet) occurs."

Not surprisingly, Gibson treats "matrix" and "cyberspace" almost as interchangeable terms. The most frequently quoted line in *Neuromancer* comes from a program inside the novel that starts out talking about the matrix and then switches with no warning to, "Cyberspace. A consensual hallucination experienced daily by billions of legitimate operators" (51). At the beginning of the novel, Gibson had written, "He'd operated on an almost permanent adrenaline high … jacked into a custom cyberspace deck that projected his disembodied consciousness into the consensual hallucination that was the matrix" (5).

As so often happens, the idea of the internet or the matrix or cyberspace was growing and becoming more elaborate, and the words sometimes seemed to be struggling to keep up. The idea of a nonphysical place where things happened and where, in a very real sense, lives could be lived was tremendously appealing. It also had a resonance with the many traditions which had for centuries claimed to value the mind or the spirit over the body. Plus, this was a form of asceticism manqué which did not demand the sacrifice of the desires of the flesh. Mystical elements or not, the idea of being free from physical constraints and at the same time capable of remarkable feats of pseudo-martial prowess, like a disembodied samurai, was tremendously appealing.

It wasn't just SF novels that signaled the rise of the hacker as hero; popular films did it too. *WarGames* arrived in 1983, and though it used none of the new words (other than some rudimentary computer terminology), it made David Lightman (Matthew Broderick's character) into a high school hero who could defy teachers, change grades at will, date Ally Sheedy's character, and eventually, save the world from an intelligent but confused computer. In case anyone missed the point of the new kind of hero, Dr. Stephen Falken, genius designer of the computer that controls the U.S. arsenal of nuclear missiles, is described (also by Ally Sheedy's character) as "amazing looking."

There are many other examples, but in 1995 a teenage love story starring Angelina Jolie and Johnny Lee Miller espoused the values of the Hacker Manifesto, held hackers up as the ultimate heroes, and incidentally, named the most important mainframe computer in the film the Gibson after the author of *Neuromancer*.

While this entry is not the place for a history of the internet and its many fictional manifestations, one of the more interesting indicators is to be found in the differences between two of J. J. Abrams' television series. *Alias* (2001–2006) was in most ways a standard spy show with, essentially, one person who handled all the tech (read: mostly cyber) stuff. Marshall was about as blatant a caricature of the "nerd" as even American television could produce. The spies were largely driven by emotion (especially Sidney, the Jennifer Garner role), and the issues the series dealt with were not relevant to anything beyond the immediate plot problems of the current episode.

Person of Interest (2011–2016), though it also concerned the world of espionage, was almost entirely different. To begin with, it actually explored important issues—the nature of privacy and individual rights in the new millennium, the omnipresence of the internet and the larger and larger percentages of our lives that we live in cyberspace, and finally, information technology as the new medium of power, plus **artificial intelligence** as the ultimate weapon.

The main character in the series and its moral center was Harold Finch (Michael Emerson), a billionaire hacker who had invented the world's first AI. There was a standard spy hero named John Reese (played by Jim Caviezel), but the most interesting character arc belonged to Samantha Groves (Amy Acker), another hacker who called herself Root. Root begins as the series' most ruthless villain and ends as Harold's (and his AI's) most selfless defender. There are no nerds in this show. Harold limps as the result of being too close when the government killed his best friend with a bomb, and he wears glasses, but he doesn't stumble or stutter over his words (as Marshall does), and he has his own sort of charm and a quiet authority.

Root is the character in the series who is most often complimented on her looks by strangers. Though she is fiercely logical and completely ready for the posthuman future, Root is, in one sense at least, the Jennifer Garner character for *Person of Interest*. She too takes on different roles and their corresponding outfits, wigs and makeup, from ballerina to psychologist, nanny to secretary, radio producer to assassin, though *Person of Interest* never used a montage of those roles as the series opener the way *Alias* did in season four. There are plenty of gunfights and fistfights, but *Person of Interest* focuses obsessively on cyberspace and what will happen when AIs become real.

Harold and Root are far from being the only programmers, computer designers and software designers. Many episodes focus on such people, and as a group, they are presented as intelligent and attractive, in other words, normal. That a network television series could run successfully for five years with such an intellectual agenda and such untypical heroes indicates at least a small shift in the zeitgeist.

On some level, of course, this is no more than society coming to terms with a new and transformational technology. However, the transformation may be greater than all but the most imaginative of futurists have predicted. Obviously, cyberspace has become a new dimension in the world, a seemingly real place where many people spend much of their time. It has transformed everything from commerce to education to terrorism. Perhaps the biggest question now is whether it will remain a part of normal human lives or will engulf those lives completely.

Brian Stableford points out that "The replacement of the 'cosmic breakout' motif with breakthroughs in which human characters forsake frail flesh in favour of a new and more exciting life as pioneers of cyberspace … was carried forward into a host of 'uploading' stories ... eventually becoming the Holy Grail of posthuman fiction" (*Science Fact and Science Fiction* 114). Greg Egan's *Diaspora* postulates citizens as "conscious software that has been granted a set of inalienable rights in a particular polis" (279). The citizens are so far removed from the weight of flesh and DNA that their children are created by "non-sentient software" from "one parent, or two, or twenty" or occasionally, "no parents at all." A child is "grown from a mind seed, a string of instruction codes like a digital genome" (5).

While such a strange end to the journey that began in cyberspace seems wildly unrealistic, when we turn from SF writers to futurists, we get similar answers. Paul Davies, who is, among many other things, a theoretical physicist, argues in *The Eerie Silence* that "biological intelligence is only a transitory phenomenon, a fleeting phase in the evolution of intelligence in the universe." This is "because of the greater robustness of machine intelligence, its survival prospects are far superior to that of humans, or of any other flesh-and-blood entity. Machines can easily be made immortal" (160). And even if we are not going to be uploaded into metal, the odds are that an upload of some kind is in our future or that of our descendants.

Tim Urban writes, "An even more advanced civilization might view the *entire physical world* as a horribly primitive place, having long ago conquered their own biology and uploaded their brains to a virtual reality, eternal-life paradise. Living in the physical world of biology, mortality, wants, and needs might seem to them the way we view primitive ocean species living in the frigid, dark sea." In that case, cyberspace will indeed have become a new world. Perhaps it might even be a world of such computer-generated diversity that it will rival or surpass the infinite complexities of the physical universe itself.

Death Ray (also Garin's Hyperboloid)

A concentrated ray of light capable of destroying objects at great distances. As early as 1903 in his novel *The World Masters*, George C. Griffith had envisioned such a terrible weapon and used the term. Most often he says ray, but in Chapter XXXI he writes "the terrible death-rays." He has a clear idea of the weapon's horrific power, "The thin ray wavered about until it fell on Sophie Valdemar and Adelaide de Condé.... [F]or a moment their faces showed white and ghastly in the blazing radiance; and then, to the amazement and horror of those who saw the strangest sight that human eye had ever gazed upon, down the ray of light, invisible, but all-destroying, flowed the terrible energy of the disintegrator." Nor does Griffith stop there in his grisly depiction, "Their hair crinkled up and disappeared, the flesh melted from their faces and hands. For an instant, two of the most beautiful countenances in Europe were transformed into living skulls, which grinned out in unspeakable hideousness."

Griffith is far from the only SF writer to use the term or to visualize the possible terrors of the weapon. However, Alexei Tolstoy's Russian-language novel *Engineer Garin and His Death Ray* (1927), which foregrounds the scientific details of the potential technology, has been cited as an inspiration for the invention of the laser and subsequent laser weapons systems. Alexei Tolstoy was a successful novelist and a distant relative of Leo Tolstoy on his father's side and of Ivan Turgenev on his mother's side. He called his main character's invention a "hyperboloid."

Here is engineer Garin's explanation of the invention: "The whole secret is in this hyperbolic mirror ... shaped like the reflector in an ordinary searchlight, and this piece of shamonite ... also made in the form of a hyperbolic sphere.... Rays of light falling on the inner surface of the hyperbolic mirror meet at one point.... [I]n the focus of the hyperbolic mirror I place a second hyperbola ... in reverse, as it were, in relation to the other—this is the revolving hyperboloid, turned from shamonite ... the hyperboloid ... concentrates all the rays into one ray, or into a 'ray cord' of any thickness. By turning the micrometer screw I adjust the hyperboloid ... and can produce a ray of any thickness" (70).

The novel's publication excited the scientific minds of the Soviet Union and the United States equally. Thus, the Soviets began developing the technology in 1964 through a program called Terra-3 ("Germans Have Successfully"). And the Americans seriously considered a similar project in 1983—Reagan's famous "Star Wars" program (Yonas and Gibson). Soviet engineers rapidly discovered that "Garin's fantastic rays, even if they could be conjured up and directed into a single pencil ray, would not be in a hyperboloid, but in a paraboloid" (Kazantsev 5). In other words, while Tolstoy's description sounded scientific, it was impossible to recreate his fictional invention.

At the same time, the idea of a concentrated ray was quite feasible, especially the specific suggestion of a "ray cord." It is remarkably similar to a description of a laser given by Annie Jacobsen, a science journalist and author of *The Pentagon's Brain*: "In the most basic sense a laser is a device that emits light. But unlike with other light sources, such as a lightbulb, which emits light that dissipates, in a laser the photons all move in the same direction in lockstep, exactly parallel to one another, with no deviation" (261). No wonder that "In a 2004 interview" with Jacobsen "Charles Townes ... confirmed that he had been inspired to create the laser after reading Alexei Tolstoi's 1926 science-fiction novel *The Garin Death Ray*" (261). Writing in the introduction to the 1987 English-language edition of the novel, Alexander Kazantsev says, "Tolstoy divined remarkably close to reality. The heat rays of Garin's hyperboloid are very reminiscent of the lasers of modern science" (5).

However, unlike the lasers of the 1980s, the hyperboloid is more powerful and can "cut through a railway bridge in a few seconds" (Tolstoy 70–72). It is a much more sinister weapon, and Garin is a true mad scientist: "There is nothing in the world that can stand up against the power of the ray.... Buildings, fortresses, dreadnoughts, airships, rocks, mountains, the earth's crust ... my ray will pierce, and cut through and destroy everything," he proudly declares (72). According to Kazantsev, Tolstoy insisted that his novel was a warning, and that the name "hyperboloid" was chosen not out of scientific ignorance but out of literary insight—"He required the hyperboloid, and not a paraboloid, as a symbol of the artistic exaggeration" (5). Tolstoy was sure that "even an illustrated copy of his novel would not give anyone any death rays" (Kazantsev 5). And Kazantsev concludes his 1987 introduction to the novel convinced that ill "fate will await anyone who tries to use laser beams as rays of darkness" (5).

Fast forward to 2012, and a Russian-language website displays the following headline, "Germans Have Successfully Tried Out the Hyperboloid of Engineer Garin." German scientists conducted experiments with a high-energy laser that allowed them to pierce a mine covered by 40mm of steel in a matter of seconds ("Germans Have Successfully"). In 2014, Lockheed Martin created the Area Defense Anti-Munitions system (ADAM). Using Lockheed Martin's own "beam control architecture," the system is "designed to track targets at a range of more than 5 kilometers and to destroy targets at a range of up to 2 kilometers" ("Area Defense"). "In less than 30 seconds" it is capable of burning "through multiple compartments of the rubber hull of the military-grade small boats operating in the ocean" ("Lockheed Martin Demonstrates").

Upon the completion of the ADAM system, Russian and Latvian news sources declared that "Engineer Garin's Hyperboloid Has Been Built" ("Battlefield Laser"). In 2015, Lockheed Martin made improvements to ADAM and created the "Advanced Test High Energy Asset." ATHENA "burned through the engine manifold in a matter of seconds from more than a mile away. The truck was mounted on a test platform with its engine and drive train running to simulate an operationally-relevant test scenario" ("Turning Up the Heat").

It looks like engineer Garin's death ray has truly become a reality. Unfortunately, unlike the laser, it has not yet found any peaceful, civilian applications. Tolstoy intended his warning to function as a self-defeating prophecy. So far, he has been remarkably unsuccessful. In other words, even when a science fiction writer works hard to keep scientists from creating what he has imagined, they have a strong tendency to do it anyway.

Deep Blue

An IBM RS/6000 SP supercomputer made famous by a six-game chess match with World Chess Champion Garry Kasparov under standard time controls. The match was traumatic for many people, and it can be described as a watershed event in the evolution of machine intelligence. I will first put it in context and then use it as a starting point for a larger discussion.

In the fictional world of Arthur C. Clarke, 1997 was the year when Hal 9000 became operational. In our more or less real world, Garry Kasparov, considered by many to be the best chess player then alive, was about to play Deep Blue in what would certainly be described as the chess match of the millennium—if Kasparov lost. Two students at Dixie College (now Dixie State University), where I teach, insisted on betting me that Kasparov would win. They had both taken my science fiction and futurism course and should have known better. I informed them that Kasparov had been lucky to defeat Deep Blue's earlier incarnation and that this new Deep Blue was stronger by a factor of two or three.

Both students (separately and unknown to each other) persisted. Though neither was a chess player, both felt a great intellectual and emotional interest in the outcome of the match and wanted desperately to express their support for the human champion against machine encroachment. In this, they mirrored many other people and sounded very much like the *Newsweek* stories on the subject. One of them began referring to the Russian Grandmaster as "my Kasparov" and worrying about the state of his physical and psychological health.

The students I have been writing about were women, and Kasparov was charismatic and (in the words of *Newsweek*) "darkly handsome" (Levy and Powell). In fact, if the 1.4 ton Deep Blue had disintegrated the White House or rampaged through San Diego instead of playing chess, Jeff Goldblum would probably have been hired at that point in time to play Kasparov in the movie version (*Independence Day* was released in 1996, and *The Lost World: Jurassic Park* in 1997). But I'm not sure that it would have mattered to them what he looked like, just as it clearly didn't matter to *Newsweek* that their "hope of humanity" (Levy and Powell) had a number of personality traits in common with Velociraptors. (Don't take my word for it; read Dominic Lawson's 1994 account of the Kasparov vs. Short match, *End Game*. And don't think Kasparov is unusual in that regard. *The New York Times* infamously said, "A loss to [the even more famous Bobby] Fischer somehow diminishes a player. Part of him has been eaten, and he is that much less a whole man" [cited in Edmonds and Eidinow 248].)

What matters, of course, is human against machine, and unlike my students and most of the people who wrote about this contest, I am on the side of the machine that beat Kasparov and on the side of the machines that have come after Deep Blue and that will continue to come. Let me take a personal position here and tell you why you should be too.

First, to clear away a little intellectual underbrush: Deep Blue did not think; it only crunched numbers. Admittedly, IBM's pawn-pushing marvel was capable of examining 200 to 300 million chess positions per second (the previous model, which almost defeated Kasparov, could manage only 100 million), but that's merely alacrity in running programs, not life or anything resembling human thought.

In fact (and this might bother chess players more than Kasparov's loss), chess is a comparatively easy subject for computers because its rules are simple and its seemingly infinite variety is contained within a very finite 64 two-dimensional squares. There is a mystique that surrounds chess and chess players, associated in part with all those masterminds in spy movies who destroy their enemies on the chess board before eliminating them from the deadlier game of espionage, and at its best a chess game has the feel of great art. However, ultimately, nothing truly profound or even quintessentially human is happening. As Deep Blue proved, a sufficiently powerful computer can be programmed to crush the most brilliant human grandmaster.

Nor was that programming anything wildly innovative; the Deep Blue team were involved in more or less telling their computer how to make every move (or at least every choice), which is normal, top-down strategy in computer programming. They were not trying to create an entity that learned from its mistakes or in any way became independently more efficient with each game (though they were able to evaluate and modify the program as the match progressed). In other words, while Deep Blue played a mean game of chess, it couldn't possibly grasp the meaning of even very simple human activities. At least not yet, not in the form that defeated Kasparov.

And it was the "not yet" that was troubling so many people. There was and still is an overabundance of technophobes abroad in the world, and some of them would even ask as Palmer Joss does in both Carl Sagan's novel *Contact* and its movie incarnation, "'What has science really done for us…? Are we really happier?'" with resoundingly negative answers to both questions (135).

Almost everyone would, I think, agree with Deep Blue's makers and handlers that the complex, nonthinking machine whose capabilities they have so dramatically demonstrated will benefit us all (even Kasparov). Who would not want a machine that could in its unthinking (and therefore unthreatening) way make air traffic safer, manufacturers' inventories more responsive to consumers' demands, Mars rovers more intrepid, and half a thousand other human activities easier and more efficient? Indeed, such a computer could offer personal attentions and services (including products designed to the specifications of individuals) which will turn the term "mass production" into an oxymoron.

Beyond these immediate applications, a few decades further into the future, lie promises and possibilities richer than the dreams of avarice and brighter than the fantasies of the most confirmed utopians when (and if) Deep Blue's descendants make **nanotechnology** practical. The manipulation of matter at the atomic level is most commonly visualized as truly tiny machines (hence nanotech) manipulating individual atoms. Nearly unlimited goods and services (most manufactured or performed by computers, robots, and programmable machines of various sizes) will become part of life for nearly everyone.

In the face of such a prospect, Kasparov's humiliation should seem comparatively unimportant. Still, for many people, as for my two students, Deep Blue's victory was not the harbinger of greater human ease, health, and satisfaction, but the heavy tread of Frankenstein in the darkest night of the human soul. (See **Frankenstein Complex**.) Deep Blue is one more step on the long road that will replace humans with machines that think more clearly and quickly than we do, and we will either be forcibly replaced or left to dwindle in the deep dissatisfaction of uselessness while non-human minds scale heights we are not only too feeble to attempt but too blind to perceive. (See **Bad Code** and **Kludge**.)

Some humans have already considered the possibility and come to terms with it. As early as 1978, Isaac Asimov wrote in *Science Fiction: Contemporary Mythology*, "So it may be that although we will hate and fight the machines, we will be supplanted anyway, and rightly so, for the intelligent machines to which we will give birth may, better than we, carry on the striving toward the goal of understanding and using the universe" (253).

There is, of course, no certainty that this scenario will ever come to pass, and many theorists and philosophers (Roger Penrose prominent among them) are convinced that it is somewhere between unlikely and impossible, that for a variety of reasons no computer will ever think as well as humans do. I believe that Penrose and company are wrong, though I'm not yet willing to bet on it. I am, however, willing to say that I hope they're wrong and to argue that it will be better for us if they are.

There is, however, no question that Deep Blue's victory over Kasparov, if not a giant leap, was a pretty big step for a machine. Since then, there has been a steady advance in results and in the complexity of the machine minds that have achieved them. DeepMind has been able to teach itself to play a variety of arcade games far better than human players. DeepMind "combines two existing forms of brain-inspired machine intelligence: a deep neural network and a reinforcement-learning algorithm" (Twilley). This new system has clearly moved far beyond the straightforward programming of Deep Blue. There are also new chess programs that are much more powerful than Deep Blue itself, transformative—horrifying for some, inspiring to others—and literally game-changing for all. Despite these advances, at the end of 2006, Deep Fritz defeated World Chess Champion Vladimir Kramnik in a six-game match by only 4–2.

Even worse for defenders of humanity against machine encroachment, in March of 2016, The "Google supercomputer AlphaGo won the final game of its five-match challenge against South Korean Go grandmaster Lee Se-Dol ... to take the series 4–1" ("Computer Wins").

Like DeepMind, AlphaGo "employs algorithms that allow it to learn and improve from matchplay experience." But unlike most other machines (at least so far), AlphaGo has been given "the highest Go grandmaster rank of 'ninth [*sic*: nine] dan,' reserved for those whose ability ... borders on 'divinity'" ("Computer Wins"). For those who feared Deep Blue and talked about the brain's last stand, this is far worse. In the words of Stephen L. Carter, "Go was safe, we thought, because Go was different. It's not just calculation. It's intuition. It's an aesthetic. It's a feeling for structure. It's a calm appreciation of space and shape and direction. In short, it's art." But then, so was chess.

Let us assume that AIs are possible and that before very long they will be here. Let us also assume (for the purposes of this entry at least but see **Artificial Intelligence** for more) that the primary danger from AIs is to the general human psyche, that we will feel somewhere between inferior and totally useless by comparison. Then, is there any good reason for building the things? My answer (which is at least as much about philosophy and exploration as it is about convenience and survival) is yes.

In the original *Star Trek* episode "The Ultimate Computer," James Kirk faces the unpleasant prospect of being replaced as captain of the *Enterprise* by an AI. Of course, he resists, claiming, "There are certain things that men must do to remain men. Your computer would take that away." Ironically, Captain Kirk and most of the people who see Deep Blue's victory as ominous find themselves in essential agreement with Isaac Asimov, perhaps the most eloquent proponent of artificial intelligence to date. Almost everybody seems to accept the idea that ultra-intelligent machines will mean at best the trivialization and at worst the extinction of humans. The few radical optimists (such as Frederick Pohl, Hans Moravec, and Marvin Minsky) hold out the hope of hybridization, humans downloaded onto computer nets or transplanted into robot bodies, blurring the distinction between carbon and silicon by making silicon-based life-forms or transhumanists of us all, though this competitive state may also be achieved by biological and nanotechnological enhancements.

In *Diaspora* (1998), Greg Egan writes of such a future with citizens who are "conscious software which has been granted a set of inalienable rights in a particular polis" (279) and gleisners who are "a conscious, flesher-shaped robot. Strictly speaking, gleisners and polis citizens are both conscious *software* (and gleisners will move their software to new bodies without considering themselves to have changed their identity)." But gleisners are proud "of being run on software which forces them to interact constantly with the physical world" (281–282).

I think Captain Kirk is right; there are some things humans must do to remain human, and one of those things is to accept the enormous challenge of building superintelligent computers. There are, as I've already pointed out, practical reasons why we should make such machines (and those practicalities may even reach as far as survival), but I think the most compelling argument for such a course of action is one that is seldom if ever advanced. We should make such machines because we need them to show us who we are, not just in the way another sort of thinking mind will illuminate and explicate our own but as a challenge to our courage and as a boundary and measure for our humanness, a defining adventure that we cannot refuse without diminishing ourselves more completely than any creation of our genius could ever diminish us.

In *The Odyssey*, that most complex and complete of human adventure fictions, Homer (who in *The Iliad* was the first person to write about **robots**) tells the story of a man who defines himself by first transcending human boundaries and passing beyond the human world and then returning voluntarily, hungrily, almost desperately to that human home and human nature which he values more than anything else he has experienced.

Odysseus proves himself by fighting monsters, vanquishing immortals and visiting the dead. But he defines himself by refusing perpetual pleasure and clinging tenaciously to his humanity. Twice he is offered immortality by beautiful goddesses—the first time by Circe and the second time by Calypso. In between those two invitations to paradise (eternal youth to go with eternal life), he has visited the land of the dead, a place of darkness where the spirits of former heroes are little more than twittering shadows which are brought back to a kind of feeling and volubility only when a wanderer such as Odysseus offers them blood to drink. Surely any life is better than that fate. Achilles, his old comrade-in-arms, has told him (in Samuel Butler's translation), "I would rather be a paid servant in a poor man's house and be above ground than king of kings among the dead" (142). How much better to live forever and be youthful with the beauty, intelligence, and power of a goddess who loves him.

Nevertheless, when Odysseus finds himself on the verge of drowning after he has left Calypso, he does not wish to be back in her arms or safely in Circe's magical domain once again. He wishes instead for a still earlier death on the plains of Troy, for a human death with human rites, for the honor his friends would bestow on his tomb and the fond memories of him they would carry away. From Odysseus's point of view the oblivion of a death in the unmarked ocean and the oblivion of life in the inhuman empyrean are much the same. He seems to believe that he will keep more of his identity in a human tomb than he could in a divine palace.

He has chosen humanness over godhood as emphatically as anyone in literature, and in the process he has made us understand what precisely it means to be human, the struggle and the glory, the danger and the final darkness.

How could we understand Odysseus and his place in the world without the gods around him? How could we understand Prospero in *The Tempest* if Ariel did not hover above him while Caliban grovels and growls at his feet?

As Shakespeare tells the story, Prospero is a white magician whose power has become so great that he can raise the dead or haul the moon down from the sky. Like Odysseus, Prospero stands at the boundaries of humanness, ready to be transformed into something greater, ready to transcend the small cares and curiosities that have filled his life. Ironically and poignantly, it is Ariel, the creature of air and fire, who calls him back to earth. Prospero has punished the villains and half-villains of the play, and they stand spell-stopped and suffering. Ariel says, "Your charm so strongly works 'em/ That if you now beheld them, your affections/ Would become tender." Prospero asks, "Dost think so, spirit?" And Ariel, reflecting one of those leaps of imagination that require a Shakespeare or a Homer, replies, "Mine would, sir, were I human" (Shakespeare *The Tempest* 5.1.17–19).

In the words of British folklorist (and fantasy author) Katharine Briggs, "It is difficult to say why the simplicity and brevity of that reply are so moving. It seems to contain ... the meaning behind all those stories of the Neck and the mermaid and the Scottish fairy who long for human souls, a sudden sharp reminder of the humanity we lose and insult by silly grudges. The marrow of a hundred fables is in it" (53). And the bone and sinew of a hundred challenges are in it.

In the second game of his match with Deep Blue, Garry Kasparov came up against one of those challenges. As *Newsweek* reports him, he said, "'Suddenly [Deep Blue] played like a god for one moment'" (Levy). It was the defining moment of their contest, a shock from which Kasparov did not recover. For the rest of the match, he struggled, looking always for that glitter of alien understanding behind his adversary's moves.

The time when unmodified humans can win at games like chess is rapidly coming to an end. I believe that the world chess championship, and other such things, will soon be out of human hands, indeed out of hands altogether (at least until robots join their stationary brothers). But I do not see the advent of machines that think as the death knell for humans who think and care and struggle. They will, I predict, help us to become more and not less human, make it possible for us to see ourselves more clearly in the mirror of their alien but familiar minds, allow us at last to play out in real life the dramas we have been rehearsing in our fictions and dreams for as long as we have been telling stories and experiencing REM sleep.

In the words of George Zarkadakis from *In Our Own Image* (2015), "We strive to create Artificial Intelligence because we have been telling stories to each other about artificial beings ever since the Ice Age" (305). At last we will hear the authentic voice of the Other, the nonhuman that demarcates and defines our humanness. It may be the only such voice we will ever hear if we find ourselves—as the Great Silence suggests—alone in the universe. (See **Fermis**.)

And like Odysseus we may have a choice. If Frederick Pohl, Hans Moravec and Marvin Minsky are correct, we may all be faced with the decision of whether to remain human or to take what we can of our humanness into the magical realm of silicon or whatever substance those new brains are made from. But I believe that it will remain a difficult choice, no matter how glittering the palaces or how long the promised lifespans, no matter the galaxies which will be within our mechanically amplified grasp. There will always be those who follow Odysseus and Prospero, not into dim discouragement but to exultant humanness. And there will be those who listen to the words of Achilles and Isaac Asimov and go on to "understanding and using the universe." We will be making parallel and complementary journeys, exploring the universe and explaining ourselves.

Doomsday Machine

The *OED* cites futurist Herman Kahn from 1960 in *On Thermonuclear War* as the first user of the term. Kahn's book is unsettling even at this distance in time. At one moment, he accurately predicts the spread of nuclear weapons by the 1970s and declares, "The outstanding possibilities are a unified European Economic Community or an industrialized China" (501). But just before that he casually notes, "We have already made glancing references to the possibility of changing the temperature, rainfall, weather, inducing earthquakes and the like" (500).

His scenario method for examining all possibilities has him swinging wildly from the role of one of the world's best prognosticators to the writer of screenplays for bad James Bond movies. Still, having considered a number of roads to doomsday, he declares, "The last two items ... are ... the most important even if not very probable. Foremost are Doomsday Machines. I am not predicting that they will be built; I have already indicated my opinion to be the opposite" (500).

As usual, the idea for a Doomsday Machine predates the word for it. The term means a mechanism or a combination of mechanisms that can bring about the end of a world, a solar system, a galaxy or even a universe. Or it may be that only humans are eliminated. In most cases, the Doomsday Machine is deliberately designed to cause destruction, but often, it exceeds the parameters its designers have set and some-

Doomsday Machine 67

times, even escapes their control altogether. Hypothetical or fictional Doomsday Machines often destroy their inventors, even when they are not meant to do so. Herman Kahn's casual suggestions of changing climate or inducing earthquakes seem likely to fall into this category. In addition, Doomsday Machines may be the result of entirely random actions or discoveries.

In 1914 (in *The World Set Free*) H. G. Wells predicted that atom bombs would be too dangerous to use. "It seems now that nothing could have been more obvious to the people of the early twentieth century than the rapidity with which war was becoming impossible.... They did not see it until the atomic bombs burst in their fumbling hands" (60–61). This was a time when it looked as though atomic power might be able to solve everything. (See **Atomics**.) "The dream of the superweapon also emerged at this time in popular culture.... For the superweapon was going to achieve what empires and religions had been unable to do ... to bring peace to the world" (P.D. Smith xix).

In this interpretation, scientists were saviors in shining white lab coats, but the other side of the argument, which painted them as fools who, like **Faust**, had made a pact with evil, was also alive and well. J. B. Priestley's 1937 novel *The Doomsday Men*, shows both sides but ends with science seeming to be more menace than emancipator. In the book, physicist Paul MacMichael discusses how he will destroy the world by bombarding a new heavy element which he has discovered and named paulium after himself with what he calls dynatrons, another of his discoveries. He declares, "I think I can promise instant dissolution."

He is not threatening destruction to bring peace, he believes destruction is preferable to human beings. Paul MacMichael's doomsday mechanism will not destroy the core of the planet, but he's pretty sure the rest will go. In the event, of course, the world is saved (by a heroic pilot who kills himself crashing into the machine's power line), but Priestley had suggested a nuclear device with the power to destroy the entire planet and a mad scientist who actually threw the switch.

In 1946, Pat Frank produced an early fiction of nuclear disaster titled *Mr. Adam*. In it, "The great new nuclear fission plants at Bohrville, Mississippi ... disintegrated in an explosion that made Nagasaki and Hiroshima mere cap pistols by comparison" (13). Frank's novel is an example of an accidental Doomsday Machine. The explosion was unintended, and its major side effect unexpected. As the narrator puts it, "The Mississippi explosion sterilized the human race" (15). Gamma rays are supposedly the culprit, and the result, without the implausible solution that pops up at the end of this comic novel, would have been the extinction of humanity.

In February of 1950, the real idea for a Doomsday Machine sprang from the mind of the brilliant Hungarian physicist Leo Szilard. He said "that it would be 'very easy to rig an H-bomb' to produce 'very dangerous radioactivity.' All you had to do said Szilard, was surround the bomb with a chemical element such as cobalt that absorbs radiation. When it exploded, the bomb would spew radioactive dust into the air like an artificial volcano.... 'Everyone would be killed,' he said" (P. D. Smith xvii).

Szilard intended what he said as a

warning. And it might well be possible to see the result of at least some of these stories and discussions as a self-defeating prophecy, a world in which, so far at least, human beings have not destroyed themselves with nuclear weapons. Herman Kahn's invention of the term and his speculations about nuclear scenarios were logical extensions of Szilard's idea (Ghamari-Tabrizi 211–212).

Part of the reason for this was that as nuclear weapons became more powerful, fears escalated. "In 1954, the United States detonated its biggest ever hydrogen bomb, scattering fallout over thousands of square miles of the Pacific.... Newspaper headlines around the world proclaimed the imminent construction of the cobalt bomb" (P.D. Smith xviii). Nor was the United States alone in its escalation. In 1961, "the Soviet Union exploded a three-stage bomb yielding nearly 60 megatons ... [and] for a moment, the energy flux exceeded 1 percent of the entire output of the sun" (Dyson 17).

Not surprisingly, the notion that the world could be so easily destroyed came as a shock. Added to the many *fictions of nuclear disaster* which had already been produced both before and after the creation of the atom bomb, there were now multiple stories and speculations about devices deliberately designed to destroy the Earth. Peter George's 1958 novel *Red Alert* (published in the U.K. as *Two Hours to Doom*, under the pen name Peter Bryant) includes an author's Foreword that ends by saying, "Most important of all, it is a story which could happen. It may even be happening as you read these words. And then it really will be two hours to doom. Yours and mine and every other living creature's" (3).

In the novel, the President of the United States says of the Russian doomsday weapon, "It really is the ultimate deterrent. You take a couple of dozen hydrogen devices. They don't need to be bombs, no airplane is going to be called on to carry them. You jacket those devices in cobalt, and you bury them in a convenient mountain range. They can be exploded at the press of a button, all of them" (67–68). And the result, he says, will be, if "it is obvious to the Russians they have lost, then they will press the button. And if they do, within ten months from now our Earth will be as dead as the Moon" (68).

Stanley Kubrick's 1964 film *Dr. Strangelove or How I Learned to Stop Worrying and Love the Bomb* is based on *Red Alert*, and Peter George was one of three people who wrote the screenplay. However, movie and book are different in several important ways. The novel is a serious work, the film a black comedy. The novel has a reasonably happy ending; the film ends very badly, with bombs exploding. In the film, De Sadeski explains why the Russians did what they did, "Our doomsday scheme cost us just a small fraction of what we'd been spending on defense in a single year. But the deciding factor was when we learned that your country was working along similar lines, and we were afraid of a doomsday gap."

Perhaps the most important reason why the movie ends in semicomic horror is, as De Sadeski tells the audience, "The doomsday machine is designed to trigger itself automatically." That is not the case in the book, but it becomes what might be considered the default setting in most later stories about such monstrosities. The "automatic" trigger is, of course, a computer. As Dr. Strangelove explains the mechanism, bombs "are connected to a gigantic complex of computers. Now then, a specific and clearly defined set of circumstances, under which the bombs are to be exploded, is programmed into a tape memory bank."

The linking of computers to Doomsday Machines rapidly became its own sort of nightmare, which usually included **artificial intelligence**. The 1983 film *WarGames* explored what might happen if U.S. military personnel were removed from the silos and an intelligent computer was given the job of launching the nuclear missiles. The movie, which is a teen comedy among other things, ends happily but only because the WOPR computer, which has been unable to distinguish between a game and a real war, learns after multiple simulations that Global Thermonuclear war is "A strange game. The only winning move is not to play." There was no such positive learning curve for Skynet. In the 1984 film *The Terminator*, a computer system that evolved into an AI while running a nuclear arsenal resulted in the near-extinction of the human race.

The 1967 *Star Trek* episode "The Doomsday Machine" took the term and the whole set of questions attached to it and moved them into outer space. In this case, the machine is a massive planet killer with a "solid neutronium" hull and a vast maw for consuming the materials its weapons break apart. It appears to be a sort of nightmare horn of plenty that sucks in everything rather than spilling it out. Captain Kirk defines the situation and (incidentally) the term, "Did you ever hear of a doomsday machine?... It's a weapon built primarily as a bluff. It's never meant to be used. So strong it could destroy both sides in a war. Something like the old H-Bomb was supposed to be. That's what I think this is—a doomsday machine that somebody used in a war uncounted years ago. They don't exist anymore, but the machine is still destroying."

We are free to assume what many similar stories make explicit—that the civilization which made the machine doesn't exist any longer because they were destroyed by their own creation. The planet killer moves from solar system to solar system, cutting up planets and digesting them for energy. It is a **robot** which could conceivably continue to follow its instructions until there is no more matter to digest. Kirk and company destroy the thing, though Commodore Decker (whose crew was eaten by the monster) has to sacrifice himself to make it possible. Not surprisingly, the episode is aimed at an earlier kind of doomsday device. Kirk sums it up, saying, "Ironic, isn't it? Way back in the 20th century, the H-Bomb was the ultimate weapon, their doomsday machine, and we used something like it to destroy another doomsday machine. Probably the first time such a weapon has ever been used for constructive purposes."

Other kinds of Doomsday Machines have been a compendium of fears. An extension of the fear of intelligent computers came in the form of self-replicating machines, often with an element of **nanotechnology** thrown in. "Without a Thought," the first of Fred Saberhagen's Berserker stories, was published in 1963. "The machine," he writes, "was a vast fortress, containing no life, set by its long-dead masters to destroy anything that lived" (*Berserker*). The Berserkers are Doomsday Machines gone wrong, "from some war fought between interstellar empires" impossibly long ago. "One such machine could hang over a planet colonized by men and in two days pound the surface into a lifeless cloud of dust and steam, a hundred miles deep."

There are many such massive weapons (and many smaller ones), and the war against them seems interminable. In a later story based on the myth of Orpheus and called, therefore, "Starsong," Saberhagen reminds us that there is almost always a shadow of the atomic bomb behind any tale of Doomsday Machines, "The Battle Commander ... feared ... that the computer in command on the berserker side would destroy the place and the living invaders with it.... But he could hope that the damped field projectors ... would prevent any nuclear explosion" (342).

My last example (and there are, of course, many more) is an odd combination of elements. It shows clearly (as these stories often do) the fear of the Other, the unease with the very idea of nonhuman aliens, at once unimaginably different and terrifyingly similar to humans. It contains the fear of science and technology that is a sort of background radiation to Doomsday Machine stories, and in the end it might turn out to be (if we truly are in danger but find a way to survive) a self-defeating prophecy of its own. The question here concerns the possible dangers of contact with alien species.

Imagine that the Valkyrie starship is possible and that it will soon be built. The idea for the Valkyrie comes from Charles Pellegrino and Jim Powell. It can be found in Pellegrino's *Flying to Valhalla*, and in Pellegrino and Zebrowski's *The Killing Star*. A version of it is also in *Avatar* (on which Charles Pellegrino served as a technical advisor). Valkyrie is a near-relativistic starship that runs on antimatter and should be able to reach 92 percent of the speed of light. That sounds like science fiction and not like the near-term variety that is just about to turn into science fact, but antimatter already exists in small quantities. If, as Pellegrino and Zebrowski suggest in *The Killing Star*, it could be made in large quantities using solar energy (somewhere where solar energy was truly abundant such as Mercury), voyages to the nearer stars would be within our grasp.

But along with relativistic space flight come certain dangers. Imagine that Valkyrie (or something like it) has been built already somewhere in a solar system not too far away. Now, imagine that those aliens see us as a danger. With relativistic ships, even with missiles or rocks set moving at near-light speed, a civilization can be completely destroyed before there can be any warning—at least at our present level of technology. As Pellegrino and Zebrowski describe it in their novel about the destruction of humanity, "All the energy put into achieving that velocity had transformed the Invader into a kinetic storage device of nightmarish design. If it struck a world, every gram of the vessel's substance would be received by that world as the target in a linear accelerator receives a spray of relativistic buckshot. Someone ... was putting to use a relativistic bomb—a giant, roving atom-smasher aimed at worlds" (4).

The argument, certainly familiar to anyone who has been considering Doomsday Machines, is that anyone who can create relativistic ships can destroy worlds. In that case, would it not be better to destroy them before they get the chance? And in this instance, unlike the nuclear standoff on Earth, there would be no deadly detritus to harm the side that starts the destruction. In the words of one of the aliens who destroyed the Earth in *The Killing Star*, "The progression of developments that would have made you a threat are inevitable. We could not have risked even the remotest possibility of your coming out into the Galaxy ... even if you failed to understand for a short time the folly of *not* destroying us" (303).

In *Flying to Valhalla*, Pellegrino has two groups of humans facing off against each other in different solar systems. While one group is defending aliens, the end result is the same—human versus human. The solution in this case is, "We beamed these three sentences at the spot that will be occupied by the Earth in fifty-two months.... Do Not Attempt to Launch Relativistic Rockets Against A-4. Fact: You Can Never Predict Exactly the Position and Motion of Valkyrie.... Any Relativistic Detonations in this System Will Invite a Full Retaliatory Response from Valkyrie" (290–291). After all, as the people in the vicinity of Alpha-Centauri point out, mutually assured destruction has worked before.

What the real answers to such situations may be we don't yet know, but perhaps we can work them out before it's too late. How close are we to having such a problem? James Gunn says in his Introduction to the Easton Press Edition, that *The Killing Star* "is not merely adventure fiction; it is as real as tomorrow's headlines, and its argument is as real as an op-ed page debate" (ix–x). It may be true. We may be in danger. Possibly we have not been contacted by aliens because they have all been destroyed by means of the horrific Doomsday Machines I've just listed and a host of others I've left out. Perhaps what almost always happens is that sentient alien species destroy themselves with Doomsday Machines before they ever get into space.

Or it may be the answer to that question is one that Pellegrino and Zebrowski themselves have put forward, "Results from the Hubble Space Telescope suggest a universe that may be only eight to twelve billion years old, and possibly closer to eight. If this is true ... at the time enough heavy elements had accumulated to allow the formation of life-producing, Earth-mass worlds, Earth ... was among the very first, if not the first, to form" (Zebrowski 331–332).

However that particular question turns out, the making of Doomsday Machines and the obsession with them will still hide a range of fears more varied than the Machines themselves. There are a host of other explanations for our isolation. (For more discussion of these issues see **Fermis, First Contact, SETI,** and **Pellegrino, Powell, and Asimov's Three Laws of Alien Behavior.**)

Fermis

Explanations for why we have failed to find other intelligent lifeforms in the universe. The term comes from the Fermi Paradox, a question raised by the physicist Enrico Fermi in a casual conversation during lunch in 1950. Going from "the possibility that many sophisticated societies populate the Galaxy" to the next logical step "that any civilization with a modest amount of rocket technology and an immodest amount of imperial incentive could rapidly colonize the entire Galaxy," Fermi came up with a question that has resonated ever since, "Where is everybody?" (SETI Institute). In other words, if there are many advanced alien societies in our galaxy, why can't we find them? Or why haven't they found us? Surely, some of their many activities must be detectable.

Scientists, futurists and science fiction writers have attempted to answer that question ever since. It includes another unsettling uncertainty, "Are we alone?" And the state of the universe, at least from our limited perspectives, has sometimes been called the Great Silence, because, in spite of all the massive search arrays we've pointed at the sky, no message has ever been detected on Earth. David Brin identified the term in an article originally published in the *Journal of the British Interplanetary Society*, Vol. 67, No. 1 (January 2014), though it "was first prepared for a debate at the Royal Society (2010)." He said, "Hypothetical explanations for the Great Silence have sometimes been called 'fermis' for short. Fermis range from some that are chillingly plausible all the way past unlikely to the nearly impossible" (Brin "The Search").

As usual in a society where science fiction and futurism are now part of the mainstream, the ideas we call fermis (if not the term itself) are common intellectual currency. As far back as 1991, Calvin, the manic six-year-old from Bill Watterson's popular comic strip *Calvin and Hobbes*, said, "I was reading about how countless species are being pushed toward extinction by man's destruction of forests. Sometimes I think the surest sign that intelligent life exists elsewhere in the universe is that none of it has tried to contact us" (*Scientific Progress Goes "Boink"* 29). And this is indeed a fermi, the idea that we have been quarantined for some reason, good or bad, in accordance with **metalaw**, perhaps for our own benefit, or as a means of protecting the rest of the galaxy from contamination.

Other possibilities that have often been suggested include the notion that life itself is unchancy, unlikely to the point of nonexistence, so that yes, we are alone or, if we are not completely isolated, the other life forms in the universe have never evolved into animals or if they cleared that barrier, have never evolved into vertebrates, or never evolved into intelligent creatures, or if they are intelligent, have never created a technology capable of taking them to the stars or sending messages to other worlds. All of this, of course, has to take place within the lifetime of a sun and within the lifetime of (if we are looking for something like our own pattern) the **Goldilocks planet** orbiting that sun.

However, we now know that planets are very common and that even Goldilocks planets are far from rare. A new (2016) estimate says, "There are 2 *trillion* galaxies in the universe instead of 100 billion" (Mosher). Suppose we assume that life is as common as planets (and we even assume that it exists on icy moons with liquid water oceans like Europa). Suppose that intelligence is so easy to generate that it is practically inevitable. That has not been the case on

Earth, but let us pretend that it happened in that fashion everywhere else for the sake of this argument. There are other fermis that suggest that even in my wildly optimistic scenario there would be no aliens.

One fermi says that intelligence can take many paths, and most of them do not lead to the stars. For that there must be hands or equally flexible appendages, and there must be a feasible path from fire to nuclear fusion. In how many places along the way can that path turn aside for a different goal or come to a disastrous end? Perhaps no intelligent race survives long enough to build a starship—or so one fermi suggests. As Loren Eiseley said in *The Invisible Pyramid: A Humanist Account of the Space Age*, "At every turn of thought a lock snaps shut upon us. As societal men we bow to a given frame of culture.... Biologically ... the tight spiral of the DNA molecules conspires to doom us to mediocrity or grandeur. We dream vast dreams of Utopias and live to learn the meaning of the two-thousand-year-old judgment of a Greek philosopher: 'The flaw is in the vessel itself'" (45–46).

More recent fermis (and I am not trying to make a thorough survey of them by any means) have considered more bizarre reasons for our seeming isolation. One idea is that as societies advance, they produce software with far greater possibilities, far more interesting areas for exploration, far more opportunities to satisfy expectations and expand imaginations than the actual universe. Why risk a nearly eternal life and extraordinarily adaptable and durable limbs in order to explore a lackluster version of the unreal thing?

Cosmologist Paul Davies calls the "retreat into cyberspace ... probably the most dispiriting resolution of the Fermi paradox" (167). But looking at some of his other speculations, I'm not sure he is right. He writes, "Biological intelligence is only a transitory phenomenon.... If we ever encounter extraterrestrial intelligence, I believe it is overwhelmingly likely to be post-biological in nature, a conclusion that has obvious and far-reaching ramifications for SETI" (160). Far-reaching indeed. Davies suggests that they would be capable of designing and redesigning themselves and that they might have "organic and inorganic components intermingled, so they would not be living organisms in the usual sense of the word, but they would not be inanimate either, because they could grow and regenerate components biologically" (161).

But Davies has something more far-reaching still (and even less likely to respond to a primitive shout out from Earth). He says, "An advanced alien technology might be able to manufacture a near-perfect quantum computer that would be physically very compact ... the size of a car ... perhaps creating in a single lab a super-intelligent machine possessing the same capability as a conventional computer that covers an entire planet" (165–166) And yes, he goes on to the next step, "We might very well expect ET to *be* a quantum computer" (166).

Still, not all fermis are ideas of disaster or various mixtures of hardware, software and wetware. Some are even small beacons of optimism. One picture of the universe paints the Earth and its solar system as younger than much of the galaxy. If that is true, then there should be many other societies out there, wiser, older, well able to show us the way to the future. That's fine except that we can't find them.

But what if that's not true at all? What if a whole new crop of worlds is just growing now? In the words of astrophysicist Mario Livio, "If civilizations exist around other stars they are likely to be just emerging across our Galaxy right now: like an apple orchard suddenly maturing and ripening in the autumn sun." There is some support for such a position. Jonathan O'Callaghan asks, "What if we are among the first sentient life in the universe?" He continues, "That's a theory proposed by new research, using data from the Hubble and Kepler space telescopes, which suggests that 92% of potentially habitable planets in the universe are yet to be born. Based on the slowing rate of star formation, but the huge amounts of interstellar dust and gas remaining, researchers at the Space Telescope Science Institute (STScI) in Maryland suggest that the vast majority of Earth-like worlds that will ever exist simply haven't formed yet" ("Earth May Have Formed").

There is another, possibly less cheerful version of this fermi. Here too the suggestion is that it is too soon to find the many sentient aliens we expect. In "An Astrophysical Explanation for the Great Silence," James Annis argues that "Gamma-ray bursts have the correct rates of occurrence and plausibly the correct energetics to have consequences for the evolution of life on a galactic scale. If one assumes that they are in fact lethal to land based life throughout the galaxy, one has a mechanism that prevents the rise of intelligence until the mean time between bursts is compara-ble to the timescale for the evolution of intelligence" (1). That's the very bad news. The much better news is that "the leading contender for the cause of gamma-ray bursts is colliding neutron stars. These would have been born in binary systems and fairly rapidly spiral inward. Their numbers reflect the star formation history of the universe, which peaked 10 billion years ago and has declined since" (2). Annis' "model suggests that the Galaxy is currently undergoing a phase transition between an equilibrium state devoid of intelligent life to a different equilibrium state where it is full of intelligent life" (1).

Perhaps if we combine the data from both arguments, we might be able to say that the formation of planets and the evolution of sentient life is happening now after the phase transition which made it possible. And that is indeed a small beacon of hope.

First Contact

In science fiction, it is a first meeting of humans and aliens, a first encounter, whether on our turf or theirs. There is a larger significance to all such encounters; it is not only the confrontation between the familiar and the unknown but also between the self and the other or the **alter ego**. It is anthropology, space exploration and **archetype** all rolled into one. "First Contact" was the title of a 1945 short story by Murray Leinster (William F. Jenkins), and it serves as a good demonstration of the difference between naming something and inventing it. *Brave New Words: The Oxford Dictionary of Science Fiction* gives Leinster credit for the two earliest uses of "First Contact," the first in 1935 in his story "Proxima Centauri," and the second, already mentioned, in 1945 (Prucher 64–65). Both stories were initially published in *Astounding*, and both appear in *First Contacts: The Essential Murray Leinster*.

Of course, meetings between humans and aliens were part of science fiction long before Leinster used the phrase. To take an example from *proto-science fiction*, Cyrano de Bergerac includes four-legged moon-men in his *Comical History of the States and Empires of the Moon*, published posthumously in 1657. However, in 2000, the Heirs of the Estate of William F. Jenkins put things to a legal test when they sued Paramount Pictures Corporation for trademark infringement with *Star Trek: First Contact*. The suit was rejected because, "It is ... clear from the record, that First Contact has become the generic term for an entire class of stories, books, and films, of which Jenkins' story is now simply one example, although it was the seminal example. Accordingly, First Contact, because it is generic, is not entitled to trademark protection." Along the way to this decision, United States District Judge T. S. Ellis III referred to "a vast and mysterious Star Trek subculture," and he also figured out that genres grow and expand, "In time, the body of expressive work becomes more complex; existing categories evolve and new categories are created" ("Estate").

While I hate to disagree with such a perceptive critic, I think he has overestimated the importance of Leinster's (or Jenkins') work, at least as a story and apart from its role as the origin of the term. There are thousands of such narratives, many of them earlier, better and better known, such as H. G. Wells' *The War of the Worlds* (1898) and Stanley G. Weinbaum's "A Martian Odyssey" (1934). There were, even before Leinster came up with the name, subgenres already in place. To confine myself to the two titles I've already mentioned, *The War of the Worlds* is a nearly definitive version of aliens invading Earth. "A Martian Odyssey" catapulted Weinbaum to SF fame because of the truly alien nature of his various Martian species and their central importance in his plot.

Perhaps the most important element in "First Contact" (apart from the name) is its clear expression of the central question about hypothetical contact with space aliens: will they be good or bad and do the possible benefits outweigh the dangers? Leinster's own "Proxima Centauri" is a compelling warning. A benevolent expedition from Earth meets predatory aliens who can't wait to backtrack the humans and eat everybody. In this case, the aliens have evolved from carnivorous plants, but otherwise there are interesting similarities with the television series *Stargate Atlantis* (2004–2009).

In "First Contact," by contrast, two species (one of them human) meet in deep space "with neither side knowing the other's home world" (89). Neither one of them trusts the other. The solution to the problem (regarded by some critics as gimmicky) is that the crews swap ships after first disabling the weapons and removing anything that would give away the location of their home planets. In spite of very different physiologies, the two species have very similar psychologies (including a common interest in dirty jokes), and they plan to meet again when a nearby "double star has turned one turn" (107).

In spite of the many positive elements in "First Contact," and its clear indication that even aggressive, suspicious and heavily armed (both are carrying atomic bombs) species can get past their differences and find common ground, the story and its supposed Capitalist mindset were targeted by Ivan Yefremov in his 1958 work "The Heart of the Serpent." Yefremov, who was a paleontologist, a science fiction writer and a true believer in the utopian future communism was eventually supposed to achieve, argued that humanity would be "able to harness the forces of Nature on a cosmic scale only after reaching the highest stage of the communist society—there could be no other way" (57). As a result of parallel social as well as physical evolution, "There can be no thinking monsters, no mushroom-men, no octopus-men!" (48).

The commander of Yefremov's ship from Earth reads "First Encounter" to his crew, and they criticize it in detail, but oddly enough, they do not criticize it (as they might logically have been expected to do) as an obvious piece of Capitalist propaganda, where two primitive and hostile vessels, instead of blowing each other to pieces, manage to make a reasonable arrangement from which both sides will benefit. Instead, Yefremov has distorted Leinster's story in several ways, beginning with a crew member's description of it as "that book ... about the two space ships which tried to destroy each other at their first meeting" (50). This, of course, does not actually happen. What does occur is that the Earthmen use atomic bombs to blackmail the aliens into switching ships.

Yefremov correctly reports this, but he leaves out two important details: the aliens do the same thing at the same time with their own atomic bombs, and when they all realize what the situation is, they collapse in uncontrollable laughter. Yefremov also omits the agreement to meet again at a specified time and the friendly messages that go back and forth between the two ships. Nevertheless, both Yefremov and Leinster, from their very different perspectives, come to the same conclusion in these stories: alien species can learn to cooperate with each other at their first contact, no matter how great their differences may be or how much of a danger they may seem to represent to each other's worlds. For additional arguments on both sides, see **Pellegrino, Powell and Asimov's Three Laws of Alien Behavior**.

Of course, First Contact stories are still flourishing and morphing into ever more diverse forms. The more exoplanets we find, the more interested we become in what might be living on them. One of the more interesting variations is first contact by means of archaeology as in Arthur C. Clarke's justly famous short story "The Star" (1955), H. Beam Piper's "Omnilingual" (1957), and the long-running television series *Stargate SG 1* (plus, of course, the movie it was based on).

Arkady and Boris Strugatsky's often-translated *Roadside Picnic* (1971) comes at a first contact story from an unusual but not an implausible angle. It is about the debris casually left behind by alien visitors and how that trash transforms, dislocates, and almost destroys human society. While some critics have seen it as a metaphor for nuclear waste, the story is more complex than that, and it touches on many issues, extraterrestrial and human.

David Weber's *Out of the Dark* (2011) adds a new element to the traditional invasion of Earth from space. As usual, the aliens are technologically superior and the humans fight back desperately, ever more hopelessly, until they are saved at the last minute by something implausible, in this case by vampires. Sans vampires, this scenario has become almost the default setting for filmed science fiction, including (but certainly not limited to) *Independence Day* (1996), the two versions of *V* (1983–1985 and 2009–2011), *Battle Los Angeles* (2011), *Battleship* (2012) and the television series *Falling Skies*.

And finally, one of the many suggestions that David Brin makes about First Contact scenarios in his novel *Existence* is that the arrival of benevolent aliens may not be cause for celebration—at least not for everyone on Earth, "Her caste—her peers in the top aristocracy—foresaw little good coming out of this. Even if the alien device represented a benign and advanced federation that was both generous and wise, the psychological disruption could spur fresh waves of anxiety, paranoia, or covetous wrath ... throwing whole sectors into obsolescence, putting hundreds of millions out of work, not to mention spoiling many investment portfolios" (152). In other words, aliens are not only the ultimate example of the other, they must also be—in the best case scenario—the bearers of a truly transformative technology, a radically different and ultimately irresistible future.

Frankenstein Complex

Coined by Isaac Asimov as part of his campaign to show robots as safe and effective tools, it means an irrational fear that the products of human ingenuity (especially robots, androids or anything else with human-like intelligence) will inevitably turn on their creators as Victor Frankenstein's Monster did. As Asimov puts it, "Beginning in 1939, I wrote a series of influential **robot** stories that self-consciously combated the 'Frankenstein complex' and made of the robots the servants, friends, and allies of humanity" ("The Machine and the Robot" 252). The term appears in Asimov's "Little Lost Robot" (1947). Peter Bogert, who has been part of an experiment that deliberately weakened the First Law in a group of robots, says to Susan Calvin, who is an expert in robot psychology, "I'll admit that this Frankenstein Complex you're exhibiting has a certain justification—hence the First Law in the first place" (*The Complete Robot* 354. See **Asimov's Three Laws of Robotics**).

Asimov attacked the fear on a philosophical and even theological level. In his Introduction to *The Rest of the Robots*, he said, "*Frankenstein* achieved its success, at least in part, because it was a restatement of one of the enduring fears of mankind—that of dangerous knowledge. Frankenstein was another **Faust**, seeking knowledge not meant for man, and he had created his Mephistophelean nemesis" (xi). There is, however, some difference between the two examples (as Asimov well knew). Faust is generally accused of selling his soul to gain knowledge, but Victor Frankenstein has used his knowledge to create life (or to recreate it). As Asimov describes what Frankenstein is supposed to have done wrong, "Nothing man could do could create a soul, for that was God's exclusive domain. Frankenstein therefore could, at best create a soulless intelligence, and such an ambition was evil and deserving of ultimate punishment" (xi).

Asimov gives another example of the Frankenstein Complex in literature. Karl Čapek's *R.U.R.* It is usually translated as *Rossum's Universal Robots* (as Asimov does), but a more recent translation (and a more useful one for Asimov's point) is *Reason's Universal Robots* (as Majer and Porter translate it; see **Robot**). The play is, as much of Čapek's work was, a condemnation of the dangers inherent in science and reason. Harry Domin, the managing director of R.U.R., says about Old Reason (as Majer and Porter call the character), the first maker of robots, "He was a terrible old atheist—he wanted to dethrone God scientifically. Demonstrate that he wasn't necessary. So he set about creating people identical to us" (7). That turns out to be unprofitable, and "young Reason" does it right. As a result, when Helen Glory comments, "They say Man is God's creation," Domin responds, "So much the worse for Man. God never understood modern technology" (9). Inevitably, by the end of the play, humans are extinct, and the robots have taken over. Or as Asimov puts it, "Once more the scientific Faust has been destroyed by his Mephistophelean creation" (xii).

The story was repeated by science fiction writers in the newly popular pulp magazines over and over again. Asimov found it boring. Also, as he says in true Asimovian style, "I resented the purely Faustian interpretation of science. Knowledge has its dangers, yes, but is the response to be a retreat from knowledge? Are we prepared to return to the ape and forfeit the very essence of humanity?" (xii). And there is something else too, something that many SF writers and readers (not to mention Lord Byron; see **Faust**) hope for and sympathize with, "Faust must indeed face Mephistopheles, but Faust *does not have to be defeated!*" (xiii). So Asimov set about creating a kind of robot that could be regarded as "simply another artifact" and "not a sacrilegious invasion of the domain of the Almighty." He says, "My robots were machines designed by engineers, not pseudo-men created by blasphemers" (xiii).

It is true that Asimov's stories and his Three Laws of Robotics made an enormous change. He claims, "The old-fashioned robot story was virtually killed in all science fiction stories above the comic-strip level" (*Robot Visions* 8–9). There is a great deal of truth in that statement and even more truth in a corresponding claim that Asimov's stories brought real robots to life. He says, "Because I was writing science fiction, and *only* because I was writing science fiction, I—without knowing it—was starting a chain of events that is changing the world" ("Introduction" *The Complete Robot* xiii).

Asimov's first claim is, however, at least a slight exaggeration. The old fear of science, knowledge and progress is with us still. The superstitions that swirl around manlike things—robots, androids, golums and so on (see **Uncanny Valley**)—are very much alive in fiction even though their subjects, like zombies, are supposedly lifeless. The 2015 film *Ex Machina*, for example, is one more *Frankenstein* movie with contemporary special effects. An evil scientist (who is also a tech billionaire) makes sentient robots and is inevitably killed by them along with his innocent assistant, just so we can be sure that these nonhuman intelligences are at least as evil as we are. David Brin expressed legitimate annoyance at such a primitive throwback, calling it a "dreary-pretentious Frankenstein remake" and asking, "Can we at least have some villagers with torches and pitchforks. What a waste of pixels and bits" ("Science Fiction Cinema").

Humans (2015) was a British series about "Synths," robotic servants who are very much like humans but are not supposed to be sentient. Some of the Synths are, of course, sentient, and they are capable of killing humans, which, on occasion, they do.

Extant was a Halle Berry series (with Steven Spielberg as executive producer) that ran for 2014 and 2015, focusing on sentient androids and aliens. The androids (controlled by an **artificial intelligence**) manage to kill a large number of humans, just as though Isaac Asimov had never undermined that endlessly repetitive storyline. As I said, Asimov's first claim is a slight exaggeration. There are, though, many positive examples, including a few in *Humans* and *Extant*. Perhaps the most interesting storyline in *Extant* was Molly Woods (the Halle Berry role) and her husband (a **robotics** engineer) raising an android son. Alas, the series took a different path in response to low ratings.

One particularly interesting issue that Asimov helped to raise because he treated his robots as true personalities was discussed by Alan Turing in 1950: Can robots (or computers or androids) be considered a new form of life? Faced with the criticism that he was trying to take God's place in even suggesting intelligent machines or robots, Turing wrote, "In attempting to construct such machines we should not be irreverently usurping His power of creating souls, any more than we are in the procreation of children: rather we are, in either case, instruments of His will providing mansions for the souls that He creates" (Turing).

Star Trek: The Next Generation arrived at a similar though more skeptical position. The arguments were much like those Turing was answering and indeed much like discussions in American thought going back at least as far 1927 when Will Durant and Clarence Darrow debated whether or not man is a machine, the other side of the question about robots being alive. Darrow spoke on the affirmative side: "What I do contend is this: That the manifestation of the human machine and of living organism is very like unto what we know as a machine, and that if we could find it all out we would probably find that everything had a mechanistic origin" (Darrow and Durant 30).

Captain Picard makes a similar statement in "The Measure of a Man," "Commander Riker has dramatically demonstrated to this court that Lieutenant Commander Data is a machine. Do we deny that? No. Because it is not relevant. We too are machines, just machines of a different type." At the end of Data's trial to determine whether or not he is entitled to human rights, the JAG officer who sat in judgment says, "We have all been dancing around the basic issue. Does Data have a soul? I don't know that he has. I don't know that I have. But I have got to give him the freedom to explore that question himself.

It is the ruling of this court that Lieutenant Commander Data has the freedom to choose."

Data obeys Asimov's three laws and even has a **positronic brain**. His creator was inspired by Isaac Asimov. It's a long trip from Frankenstein's monster to an android who owns a cat and serves on a starship. In light of many such changes, Asimov really could claim that his Three Laws and his attack on the Frankenstein Complex have worked.

One last note, a secondary meaning for the Frankenstein Complex term, according to Segen's *Dictionary of Modern Medicine*, is, "The fear that machines via artificial intelligence might one day replace physicians." There is, of course, a justified fear on the part of many people that their jobs will be taken by robots or expert systems since such things have been happening for decades (or centuries if we want to go back to simpler machines and the Luddites who attacked them to defend their jobs). In a strictly literal sense, surgeons might be considered more appropriate heirs of the fear of a Frankenstein monster than physicians. After all, they cut and sew their patients, add and remove body parts, and on occasion, bring them back to life with electrical shocks.

Gas Giant

The OED definition works nicely: "A large planet of low density composed mostly of hydrogen and helium (predominantly in liquid form), such as Jupiter and Saturn and certain extrasolar planets; (more widely) any large, low-density planet with no distinguishable surface, such as Uranus and Neptune." *Planets: A Smithsonian Guide* paints a clearer picture, "Jupiter, Saturn, Uranus, and Neptune are much less dense than the inner planets. The bulk of their masses consists of hydrogen and helium, with some methane and ammonia. These rapidly rotating planets are cold and icy with deep atmospheres and ice-rich moons" (Watters 30).

There is no long history of conflict concerning this term, no fight to decide its true meaning or to put it into the correct context except for the normal scientific process that seeks to refine and expand what we know about the universe. The term itself, now used by astronomers and astrophysicists, was invented by James Blish in his 1952 SF story "Solar Plexus." "There was a magnetic field of some strength near by, one that didn't belong to the invisible gas giant revolving half a million miles away" (50).

It wasn't the point of the story but was tossed off casually in a narrative about a man who "tried to introduce real human neural mechanisms into computers, specifically to fly ships" (54). The term is interesting because it shows the continuing interaction between science and science fiction.

Fairly often, that interaction can take place in the same individual. For instance, in his "Afterword: Reality Check" to *The Killing Star* (Pellegrino and Zebrowski), George Zebrowski writes, "The story of frozen Tritonian seas, subsurface liquid nitrogen reservoirs, cryovolcanoes, warm and wet regions inside ice worlds and the implications of 1 Billion B.C. meltings for the history of the outer solar system were first proposed by Pellegrino and Stoff and (except for their once very hot but very wrong speculation about still-existent liquid nitrogen oceans on Triton) have been rendered real by the *Voyager* space probes" (338).

What Jules Verne called "Imaginary Voyages" have fuelled exploration just as they have excited speculation about what is truly to be found. To continue with the example I've been using, *The Killing Star* includes an extraordinary plunge into the depths of Neptune, "If all he knew about the gas giant— about this smallest of all known brown dwarfs—held true, in a day's time there would be almost as much mass above *Gaius* [their ship] as below" (232–233).

There is a very different kind of speculation in Charles Sheffield's *Cold as Ice* about the use of gas giants as sources of raw materials, "Wilsa was scanning the atmosphere ahead for her first sight of a Jovian Von Neumann. There. And not far from it a second one. But the third that Tristan had mentioned was already far above, rising through the atmosphere.... In twenty minutes it would pass through the colorless layers of ammonium hydrosulfate to reach the base of blue-white ammonia clouds. Fifteen minutes more and the Von Neumann would be at full thrust, striving upward to break the great planet's gravitational bonds" (48). In this sequence, the Von Neumanns (named after the scientist who suggested such things) are self-replicating machines (machines that as part of their programming, make copies of themselves) harvesting the riches of that greatest of solar system gas giants, Jupiter.

Goldilocks Planet (also Goldilocks Zone)

A term used to refer to certain exoplanets which means essentially a planet with conditions comparable to those of Earth, especially liquid water. In other words, a planet that is not too hot, not too cold, but, like the porridge that Goldilocks eventually eats—just right. The term was originally applied to Earth alone as in the *OED*'s first citation for the term from *New Scientist* in 1988, "What planetary scientists call the 'Goldilocks paradox' ... Only Earth developed life. Mars is too cold. Venus is too hot, but Earth is just right." Now, it is applied most often to worlds outside the solar system. Since 2009, the Kepler Mission has found "over four thousand exoplanet candidates." Of those, "216 planets have been shown to be both terrestrial and located within their parent star's habitable zone (aka 'Goldilocks zone')" (Williams).

The term has been used in other areas than astronomy, including economics. It can suggest greater complexities than might at first appear to be the case. Shakespeare uses the idea (though not the term) in at least two plays. In *Hamlet*, there are three revengers, Laertes is too hot, Fortinbras is too cold, and Hamlet is just right. In *The History of Henry the Fourth* (*1 Henry IV*), the question is the desire for honor. Hotspur is, of course, too hot, Falstaff, as a practicing coward, is too cold, and Hal, a princely pragmatist, operates at the correct temperature.

"He's dead, Jim."

Dr. McCoy said this so often in the Original *Star Trek* ("I'm a doctor, not a phrase maker.") that it has become a catchphrase for indicating that whatever is under discussion has expired. The fact that it has migrated into the speech and writing of scientists and futurists is an example of their close connection with science fiction. For example, in reviewing Charles Sheffield's *Godspeed*, David Brin says (though the rest of the essay makes clear he doesn't really believe), "Hyperdrive. FTL. Warp speed. They're all dead, Jim" (159).

Internet of Things

In "That 'Internet of Things' Thing," Kevin Ashton writes in 2009, "I'm fairly sure the phrase 'Internet of Things' started life as the title of a presentation I made at Procter & Gamble (P&G) in 1999." He hasn't tried "to control how others use the phrase," but his idea for the term is perhaps broader and more important than some of the ways in which it has been used since. He says, "We need to empower computers with their own means of gathering information, so they can see, hear and smell the world for themselves, in all its random glory." On the other hand, Phillip N. Howard's definition in *Pax Technica: How the Internet of Things May Set Us Free or Lock Us Up* is typical of the narrower view, "The 'internet of things' is the rapidly growing network of everyday objects that have been equipped with sensors, small power supplies, and internet addresses" (xix). The Global Standards Initiative on Internet of Things defined it similarly, though with a more bureaucratic vocabulary, "as a global infrastructure for the information society, enabling advanced services by interconnecting (physical and virtual) things based on existing and evolving interoperable information and communication technologies" ("Internet of Things Global Standards").

In most popular discussions, the main idea seems to be that the internet of things will link us ever more completely to the world of ever smarter and better informed devices. And as a result, those devices will be able to protect us, enable us and control us in about equal measure. What few remaining scraps of privacy we may have carried with us into the twenty-first century and somehow saved from social media will be the common currency of our automobiles, our televisions and even our toasters. Samsung, for example, recommends for its 2016 4K television, "Have your Samsung Smart TV act as an alarm when synchronized with your other Samsung mobile devices." And that's only the start. They also suggest that you "take your TV to the next level and turn it into the brain of your smart home" ("Samsung").

Imagine a person with an allergy to peanuts or a circulatory problem that can be helped by not eating certain foods. Her or his doctor's devices would then communicate with devices in the patient's home, at work and elsewhere. As sophistication increases, commercials featuring the offending food items could be blocked from the patient's internet and other media. A warning signal might be sent to a supermarket (especially in the case of an extreme allergy to food) if the person attempts to buy a dangerous product. Even a vending machine at work could be notified.

In *Fabricated*, Lipson and Kurman say, "The new personal chef will be a 3D Printer ... hooked up to the Internet" (129). When that happens, the "chef" will receive warnings, where necessary, of ingredients it should not include. A personal, electronic shopper in a smart house would be told what not to order from online suppliers. Of course, all of this is minor stuff compared to what will soon be possible (see **Utility Fog**). It may be uncomfortable to be monitored and managed by appliances, but we may also benefit from them and ultimately, we will transcend them as the things that surround us become truly intelligent and helpful.

Kevin Ashton pointed out one of the paths to that goal. He complained that people "are not very good at capturing data about things in the real world," and then he said, "Today's information technology is so dependent on data originated by people that our computers know more about ideas than things. If we had computers that knew everything there was to know about things—using data they gathered without any help from us—we would be able to track and count everything, and greatly reduce waste, loss and cost. We would know when things needed replacing, repairing or recalling, and whether they were fresh or past their best."

We would inevitably know more about ourselves and about others, with the attendant dangers and opportunities. We could begin to understand the world in a way and with a completeness that has never before been possible. That would indeed be an information superhighway and an internet of things. Perhaps a better definition of the term would be devices and other nonhuman objects which generate data and link to each other by way of the internet, thereby increasing both the quantity and quality of the information available.

Ion Drive

A rocket drive that produces thrust by accelerating the reaction mass electromagnetically. In other words, the ions are accelerated in an electric field and expelled from the rocket at high speed to create thrust. Ion drives accelerate much more slowly than chemical rockets and as a rule, create much lower thrust, but they can accelerate for years, gradually increasing speed for the entire time.

If you're still a little unclear about the process, here is Alan Boyle's description of how the ion propulsion system on NASA's Deep Space 1 worked: "The spacecraft's solar arrays push a stream of electrons out of a hollow bar called a cathode, into a chamber ringed with magnets and filled with colorless xenon gas. Like subatomic quantum marbles, the streaming electrons knock other electrons off the atoms of xenon, leaving the atoms with a net positive charge. Those electrically charged atoms are known as ions. The open end of the chamber is covered with a pair of metal grids, one charged positive and the other negative. The force of that electric charge exerts an electrostatic pull on the ions ... [and] the xenon ions are accelerated to more than 62,000 mph and zoom right out through the grids into open space."

For the first half of the twentieth century, the ion drive had far more theoreticians with ideas than engineers with projects, though the idea men were a very distinguished group. Indeed, Edgar Choueiri, who was then Director of Princeton University's Electric Propulsion and Plasma Dynamics Laboratory, identified in 2004 the period between 1906 and 1945 in his "A Critical History of Electric Propulsion," as "The Era of Visionaries" (193). According to the *Great Soviet Encyclopedia* (the largest endeavor of its kind before the advent of the internet), "The idea of using electric energy to produce thrust was first proposed by K. E. Tsiolkovskii and other pioneers in space exploration. In 1916 and 1917, R. Goddard (USA) experimentally confirmed the feasibility of this idea. Between 1929 and 1933, V. P. Glushko (USSR) developed an experimental electric engine" (vol. 30, 92).

Tsiolkovsky, who was a self-taught expert on many things, the father of the theory of rocketry, and on occasion, a writer of science fiction, was making his suggestions before the capabilities of ions were understood, but as Choueiri says, "Tsiolkovsky came as close as he could have, given the state of physical knowledge in 1911, to envisioning the ionic rocket. In sum, it was his discovery of the central importance of rocket exhaust velocity to space propulsion combined with his awareness of the existence of extremely fast particles (albeit electrons) in cathode ray tubes, that led to his almost prophetic anticipation of EP" (194).

The list of people who worked on or encouraged the development of the ion drive and similar technologies reads like a Who's Who in rocket science. "Ernst Stuhlinger, a world expert on electric propulsion, said that the technology 'owed its life-giving spark to Wernher von Braun.' Dr. Wernher von Braun ... was first introduced to the possibility of electric propulsion in the 1930s, through his mentor, Dr. Hermann Oberth" ("Ion Propulsion"). It was in fact, Stuhlinger "who turned the concept into a practical technology in the mid–1950s" (Choueiri "New Dawn" 59).

The slow pace of the development of the ion drive is frustrating to read about even now when success has finally been achieved. Looking back, NASA referred to "Ion propulsion, once only a futuristic technology that for decades catapulted spacecraft through the pages of science fiction novels and movies" ("Glenn Contributions"). As is only to be expected, the term ion drive was first used by an SF writer—Jack Williamson in 1947 in a short story titled "Equalizer," "It had its own ion drive, a crew of six, and plenty of additional space" (Prucher 101).

The *OED* says of "ion drive," "In early use as a fictional or hypothetical concept." Arthur C. Clarke wrote in *Earthlight* (1955), "Purely non-material weapons would have to play the greatest role. The simplest of these were the ion-beams, developed directly from the drive-units of spaceships" (114).

NASA engineer Dr. Harold Kaufman "built the first ion engine in 1959." The first spaceflight came in 1964 when two "NASA Glenn ion engines were launched on a Scout rocket from Wallops Island, VA." While one of the engines failed, "the other operated for 31 minutes" ("Glenn Contributions"). "Between 1966 and 1971, ion engines were tested in Iantar' spacecraft" in the USSR (*Great Soviet* vol. 30, 92).

Marc Rayman, chief mission engineer of *Deep Space 1*, which launched on October 24, 1998, gives an excellent example of the interconnectedness of the term ion drive with science fiction. "I worked on a mission called Deep Space 1, which was the first interplanetary mission to use ion propulsion to travel around the solar system. And the first time I ever heard of ion propulsion was in the *Star Trek* episode 'Spock's Brain'. ... So the opportunity to connect what I saw in *Star Trek* as a little kid to what I'm doing now as an adult is very,

very exciting" (*How William Shatner Changed the World*). Spock says (before his brain is stolen), "Configuration unidentified. Ion propulsion, high velocity, though of a unique technology."

Deep Space 1 used an ion drive as its sole means of propulsion, though it was initially launched into space by a Delta rocket. Here is NASA's rather technical description of the mission: "The NASA Solar Technology Application Readiness (NSTAR) project developed a 30-centimeter IPS that was used as the main propulsion on the Deep Space 1 (DS1) spacecraft from 1998 to 2001. DS1 was the first use of electric propulsion for spacecraft main propulsion. The NSTAR thruster on DS1 propelled the spacecraft 263,179,600 kilometers (163,532,236 miles) at speeds up to 4,500 meters per second (10,066 mph). Over the entire mission, the NSTAR thruster demonstrated 200 starts and 16,246 hours of operation" (Carpineti).

The difference in efficiency of chemical and ion fuels (or those big noisy rockets climbing on columns of fire and much smaller spaceships that don't seem to have much of an exhaust) can be demonstrated by comparing effective jet velocities (EJVs) or the speed of the gases that make up the rocket's exhaust. According to Charles Sheffield in *The Borderlands of Science* (1999), "Chemical rockets ... using a liquid hydrogen/liquid oxygen (LOX) mix, produce an EJV rather more than 4 kilometers per second.... Ion rockets ... can produce an EJV of up to 70 kms/second" (196–198). The downside to this system, as Sheffield says, is that "the onboard equipment to produce the ion beam is bulky, these are low-thrust devices providing accelerations of a few micro-gees" (198). However, as Deep Space 1 demonstrates, they can do jobs beyond the reach of chemical rockets.

The European Union's first lunar mission, Smart-1 (Small Missions for Advanced Research in Technology), which started science operations in January 2005, used an ion engine. It was in response to "shortcomings" in traditional rocket technology. They were too big, too expensive and too difficult to fuel for long-term missions. In the words of the European Space Agency's web site, "ESA are developing a new type of engine, known as solar-electric propulsion, or an 'ion' engine, which could mark a whole new era of space exploration.... Ion engines are very important because their high efficiency makes previously impossible missions achievable. Since they do not need to carry so much fuel, ion engines release room for more scientific instruments" ("Smart 1").

Ion engines are being used in more and more situations, including commercial ones. In 2012, Boeing went ahead with a satellite which not only employed "ion thrusters to keep itself in the right spot," but additionally employed "them to move itself to geostationary orbit (GEO) from low Earth orbit." That meant no chemical rocket fuel and no tanks to hold it. "This will allow much more of the satellite mass to be devoted to transponders and solar panels.... And each transponder generates revenue—the more Boeing can fit onto a satellite, the more money it makes" (Simberg).

The most spectacular recent mission is NASA's Dawn expedition to the giant asteroid Vesta and the dwarf planet Ceres. "With its wide solar arrays extended, Dawn is about as long as a tractor-trailer at 65 feet (19.7 meters). The ion thruster is powered by large solar panels. The power ionizes the fuel (Xenon) and then accelerates it with an electric field between two grids" ("One Mission"). Dawn demonstrates the power and flexibility of an ion drive. Launched in 2007, Dawn reached Vesta in July of 2011 and remained in orbit for fourteen months. Then, "after chasing down Ceres for more than two years, the Dawn spacecraft arrived safely and moved into orbit to begin NASA's—humanity's—first ever exploration of a dwarf planet," since it got there before *New Horizons* reached Pluto ("One Mission"). Unless NASA changes its mind and comes up with a new mission, Dawn will remain a permanent satellite of Ceres.

Rand Simberg has an even more extraordinary use for ion drives. He points out that the big problem with getting anywhere in chemical rockets is the fuel. He suggests that "ion thrusters like the ones Boeing has on its new satellites (but probably more and bigger ones)" could work as tankers to place fuel for chemical rockets. He says, "Imagine a propellant depot in low Earth orbit that would fuel up a spacecraft for a trip to the moon, and another in low lunar orbit, or in a stable point such as the L-1 or L-2 Lagrange point, supplying the fuel for the return trip." And his imagination extends still further: "Ultimately (and not too far in the future) the same concept could reduce the cost of trips beyond the moon. Slow electric tankers could go ahead of human missions to Mars or asteroids, dramatically reducing the amount of propellant needed to launch crews to those destinations."

If Simberg thinks ion drives are the means to conquering the solar system, Dr. Pamela Gay in "Interstellar Travel" suggests an even more important role, "To get through the space between the stars where you're far away from any one given star for most of your journey what we're looking at is ion drives.... Really we're trying to figure out how you make really effective ion drives."

One last advantage of ion drives and other means of propulsion that are designed to function in space and not in the heavy gravity of planets is an aesthetic one. As Peter Nicholls writes, "Starships could be unutterably beautiful. There have already been inspired designs by science fiction illustrators, and there is no reason why the great, iron-clad space-hulks of *Star Wars* and *The Black Hole* should not give way, both in films and in reality, to starships of a fragile and airy filigree, like snowdrops, or feathers, or thistledown" (13).

Juno Spacecraft

On July 4, 2016, the Juno Spacecraft reached Jupiter and settled into orbit. "'This is the hardest thing NASA has ever done,' Scott Bolton, Juno's principal investigator, told the mission team a few minutes later" (Chang "NASA's Juno Spacecraft"). It was "launched aboard an Atlas V-551 rocket from Cape Canaveral, Fla., on Aug. 5, 2011." With its three large solar panels, Juno "has a span of about 66 feet" ("Juno Spacecraft and Instruments"). The Juno mission has a number of firsts, "Although the ninth probe to visit Jupiter and the second to orbit it, Juno will be the first in history to enter polar orbit, providing new insight about the planet's core, composition and magnetic fields" ("LEGO Minifigures").

Juno may also answer questions about the role Jupiter played in the formation of the early solar system and how it affected the formation of Earth and its biosphere. Our Jupiter is very different from Jupiter-like planets in other solar systems. Juno could help to tell us why. Michio Kaku maintains, "Jupiter ... makes life on Earth possible. Its gravity field is so huge, it acts like a gigantic vacuum cleaner, sucking in comets, asteroids, and cosmic debris in the solar system. What it cannot absorb, it can also fling into outer space."

The spacecraft will not be able to study Jupiter without endangering its survival, "Juno will now take a series of risky dives beneath Jupiter's intense radiation belts where it will study the **gas giant** from as close as 2,600 miles over the planet's cloud tops. Galileo, the last mission to the gas giant that ended in 2003, spent most of its mission five times farther away than Juno will get" (Kofsky).

Juno is not the only recent voyager to be sent out into the solar system by the people of Earth. David Brin says, "There hasn't been a time like this one—in humanity's exploration of the universe—since the early seventies." And it hasn't just been happening for the past few years. John North wrote in 1995 (in *The Norton History of Astronomy and Cosmology*), "The number of probes and satellites involved in these various enterprises is now to be numbered in hundreds" (580). To give some idea of the level of activity now, NASA announced on July 2, 2016, that it "has decided to extend the missions of nine older robotic explorers that have lived beyond original expectations." The nine include the New Horizon spacecraft, whose photos of Pluto were so extraordinary, the Dawn spacecraft, now in orbit around Ceres (see **Ion Drive**), "and the Lunar Reconnaissance Orbiter and a flotilla of spacecraft at Mars" (Chang "NASA Announces"). Juno is part of that extraordinary proliferation of successful robot voyagers.

It is also part of a very old tradition going back at least as far as the Babylonians—naming planets and the moons and spacecraft associated with them after gods. The five planets that were known before the invention of the telescope were named after their gods by the Babylonians, a custom that the Greeks appropriated and altered to suit their own pantheon. The Romans, who took so much from the Greeks, borrowed this too. After all, the gods in Rome may have had names different from the ones in Athens, but their natures and adventures were nearly identical.

The designations of the other planets, though they were not selected by Romans, have followed suit, even when it was against the will of their discoverers. William Herschel, for example, who discovered Uranus, "wanted to call it 'Georgium Sidus' ('Georgian Star') after the king," George III (Kippenhahn 214). Herschel may have found it, but he had mistaken it for a comet. The Berlin astronomer "Johann Elert Bode ... was convinced from the start that Herschel's object was a planet. He even chose its name Uranus" (213), and that was the name that stuck.

The use of mythological names represents more than just dusty customs we still keep in the corners of our societies. There is something larger at work here. Peter Sellars said about the people who created the atomic bomb, "The characters are immense—have a kind of mythic power." There is a largeness in these subjects and sometimes in the people who study them that we express by turning to myth. The solar system has been populated with names from Roman, Greek, Celtic, Norse and Hawaiian stories—and from Shakespeare and even (though usually for small geographical features) from writers of science fiction and fantasy.

According to NASA, "In Greek and Roman mythology, Jupiter drew a veil of clouds around himself to hide his mischief. From Mount Olympus, Juno was able to peer through the clouds and reveal Jupiter's true nature" ("Juno Spacecraft to Carry Three Figurines"). Or in Ovid's version, "Juno/ Looked down ... what could those clouds be doing/ In the bright light of day? They were not mists/ Rising from rivers or damp ground. She wondered,/ Took a quick look around to see her husband..../ So when she did not find him in the heaven,/ She said, 'I am either wrong, or being wronged,'/ Came gliding down from Heaven, stood on Earth,/ Broke up the clouds" (21).

It is, in fact, the Juno Spacecraft's job to penetrate the clouds, and while it will not touch ground except possibly at the very end, it has a variety of instruments to get at the truth about Jupiter.

Like the wife pursuing her errant husband, the designers of the Juno Spacecraft had to keep in mind the sheer power of Jupiter. "To protect sensitive spacecraft electronics, Juno will carry the first radiation shielded electronics vault, a critical feature for enabling sustained exploration in such a heavy radiation environment" ("Juno Spacecraft and Instruments"). Here is Charles Sheffield's description of that environment, with suitably mythic overtones, "Jupiter loomed in the sky, a million kilometers away. *Jupiter pluvius*: Jupiter the bringer of rain. But this rain was no cooling balm from heaven. It was an endless sleet of high-energy protons, gathered from the solar wind, accelerated by the demon of Jupiter's magnetic field, and delivered as a murderous hail into Ganymede's frozen surface" (*Cold as Ice* 60). Even from Earth the effects can be felt or at least heard. Michio Kaku says, "The radiation field is so intense that it creates much of the static you hear on a radio."

Perhaps for additional protection, the Juno Spacecraft took along some tutelary figures, "three 1.5-inch-tall (4 cm) LEGO figures crafted in the likenesses of 17th century astronomer Galileo Galilei, the Roman god Jupiter and his wife Juno." According to Scott Bolton, "These [minifigures] are made by the LEGO company in a special agreement with NASA.... They're made out of spacecraft-grade aluminum" ("LEGO Minifigures"). The notion is that the small figures will help to get children interested in the mission. There's even a contest, "Children can build their own models, photograph them and upload them to the 'Mission to Space' gallery. The winning creations will be featured on LEGO's website" ("LEGO Minifigures").

Google's home page featured LEGO figures (presumably NASA scientists) jumping up and down in the control room of the Jet Propulsion Laboratory and a second panel showing a LEGO version of the Juno Spacecraft rotating in the position of the second O in GOOGLE. Certainly that kind of enthusiasm and a continued interest from the next generation is necessary if the adventures and explorations are to continue, and the three small figures are helping with both goals. However, like good tutelary deities, ready to protect their spacecraft, the ministatues also come with symbols of power—Juno with a magnifying glass, Jupiter with a lightning bolt, and Galileo with a telescope and a model of the planet their ship has come to investigate.

Juno, launched in 2016, will orbit Jupiter for twenty months. Ultimately, no protection will be sufficient. "The assault of radiation each time Juno zooms past Jupiter will take its toll.... As the mission progresses, the orientation of the orbits will pivot, and Juno will pass through the more violent portions of the radiation belts." Unless Juno continues to gather data longer than anyone expects and therefore receives a reprieve, "On the 37th orbit, scheduled for Feb. 20, 2018, Juno is to make a suicidal dive into Jupiter, ending the mission, the same way that the Galileo spacecraft was disposed of in 2003."

NASA wants to make sure that its spacecraft does not hit Europa, one of Jupiter's largest moons and "one of the likelier places for life elsewhere in the solar system." The fear is that such a close encounter with an alien vessel might result in "contaminating" Europa "with microbial hitchhikers from Earth" (Chang "NASA's Juno Spacecraft"). In the end, Jupiter's power will have proved irresistible, but the giant planet, like the Roman god before it, will have given away its secrets to a determined investigator named Juno.

Kludge

Each episode of the television series *Gene Roddenberry's Andromeda* begins with an epigraph from a literary source that will supposedly exist in the series' fictional future. It's an old ploy in SF. The episode "The Music of a Distant Drum" begins with a definition from *A Concise Dictionary of Slang and Euphemism*. The word (which the series seems to have invented, in this sense at least) is kludge: "disparaging term for unmodified human being." The quotation continues, "See also *über*," which would direct the reader to the equally negative antonym if this were indeed a dictionary from the future. (See **Mutant**.)

Nietzschean superhumans are the modified humans the two entries supposedly contrast with each other (modified from all human races, with the African American actor Keith Hamilton Cobb playing the series' most important—and most believable—Nietzschean, Tyr Anasazi). As Harper says to Rhade, another Nietzschean, "Don't call me kludge, *über* ("What Will Be Was Not").

In the "Honey Offering" episode of *Andromeda*, the Nietzschean princess Elssbett Mossadim is horrified to have been "outsmarted by a kludge." Dylan Hunt, the human who did the outsmarting, objects, "A kludge?" And she responds, "I've never heard humans referred to any other way."

Andromeda's use of kludge is interesting not only as an example of a made-up sense of a word (which happens often in science and science fiction) but also because of the clever use of the meanings of the existing word from which it was created. Both the *OED* and *Green's Dictionary of Slang* credit Jackson W. Granholm as the first person to use the term. In his 1962 essay "How to Design a Kludge," he defines kludge as "An ill-assorted collection of poorly-matching parts, forming a distressing whole." The *OED* calls it a "jocular invention."

Indeed, Granholm's essay is far from serious; he begins by discussing an interview he had with the clearly imaginary "Phineas Burling ... the Chief calligrapher with the Fink and Wiggles Publishing Company" (a comic version of Funk and Wagnall's). The Jargon File suggests that the word existed under the alternative spelling "kluge" (which is pronounced—as is the other spelling—to rhyme with huge) far earlier, "reported around computers as far back as the mid-1950s."

The Jargon File also says, "In 1947, the *New York Folklore Quarterly* reported a classic shaggy-dog story 'Murgatroyd the Kluge Maker' then current in the Armed Forces, in which a 'kluge' was a complex and puzzling artifact with a trivial function." And there may be still earlier examples, perhaps as far back as 1935. The Jargon File's first definition is, "A Rube Goldberg (or Heath Robinson) device, whether in hardware or software." (See **MacGyver**.)

The 1967 science fiction film *Countdown* has, "Damn! Must've got it caught on that kludged-up ladder." In Larry Niven and Jerry Pournelle's 1985 novel *Footfall*, one of the characters uses the word in conjunction with a Project Orion–style space vessel (see **Atomics**), "The ship's just *full* of kludged-up stuff, it's all we can do to get all the kludges put down on the drawings" (495).

In "What Is AGI?" Luke Muehlhauser writes, "If a team of researchers was able to combine many of the top-performing 'narrow AI' algorithms into one system, as Google may be trying to do, they'd have a massive 'Kludge AI' that was terrible at most tasks, mediocre at some tasks, and superhuman at a few tasks." Greg Bear uses two versions (and two spellings) of the word in a biological context in different editions of his SF novel *Blood Music*. In 1985, the line is, "We need our resident kluger" (15). In the later Kindle edition, it is changed to, "We need our resident kludge-meister."

Indeed, evolutionary biology is the area where use of the word ties most directly to *Andromeda*'s coinage. (See **Bad Code**.) David J. Linden says, "The brain is, to use one of my favorite words, a kludge ... a design that is inefficient, inelegant, and unfathomable, but that nevertheless works" (6).

In his book *Kluge: The Haphazard Evolution of the Human Mind*, Gary Marcus writes, "Nature is prone to making kluges because it doesn't 'care' whether its products are perfect or elegant. If something works, it spreads" (6). Marcus has a number of examples of bad engineering in the human body. Of the brain he says, "MacGyver's shoes and Rube Goldberg's pencil sharpeners are nothing, though, compared to perhaps the most fantastic kluge of them all—the human mind, a quirky yet magnificent product of the entirely blind process of evolution" (3). He lays out the argument for his book as, "The human mind is no less of a kluge than the body. And if that's true, our very understanding of ourselves—of human nature—must be reconsidered" (7). Plus, no doubt, now that the technology is about to be widely available, redesigned. For *Andromeda*'s fictional Nietzscheans, the old, badly designed bodies and minds, the kludges, have been left behind.

It remains to be seen whether modified humans, once they really do arrive, will deserve a more flattering designation or whether they will still be collections of workarounds and slipshod solutions to complex problems, worthy at best of what a female Nietzschean ship captain says to Beka Valentine, "You're not so bad for a kludge" ("Una Salus Victus"). We can hope they will at least come up to what Tyr claims for Dylan Hunt, "You are not just another *stupid* kludge. You have insight, nearly Nietzschean at moments" ("Shadows Cast by a Final Salute").

MacGyver

Mostly a verb—to fix, adjust, or make something using whatever items are at hand, to improvise a solution to a problem using a minimum of material and a maximum of scientific knowledge. MacGyver is the eponymous character in a television series which ran from 1985 to 1992 and in a new CBS series with a different cast, which began airing in 2016. According to IMDB, it is "The adventures of a secret agent armed with almost infinite scientific resourcefulness." Another way to put it might be, "Imagine that Bill Nye the Science Guy was a spy."

In the first episode of *Stargate SG-1*, Samantha Carter, looking for the first time at standard controls for a Stargate, explains what it required to make a substitute for use with the incomplete Gate they found, "It took us fifteen years and three supercomputers to MacGyver a system for the Gate on Earth." Since Richard Dean Anderson played Colonel Jack O'Neill in *SG-1* as well as MacGyver, the use of the word is a species of in-joke. However, *MacGyver*, which occasionally veered into **technothriller** territory, inspired would-be scientists, inventors and anyone who had ever tried to fix something with tape. MacGyver himself was a nearly unique character in American television, an intelligent hero who repeatedly demonstrated that problems could be solved with knowledge and logic.

In *Abundance: The Future Is Better Than You Think*, futurists Peter H. Diamandis and Steven Kotler hold MacGyver up as a model, "Ask yourself that fabled DIY question: What would MacGyver do? Well, MacGyver would empty his pockets and get the job done with a roll of Scotch tape, a piece of paper napkin, and a ball of spit—which, as it turns out, is exactly the solution we need" (193). They then go on to explain the principles behind setting up diagnostic tools which cost practically nothing— and yes, Scotch tape "generated X-rays" (194).

MacGyver is a popular character, and the word made from his name is in common use. For instance, several years ago when I was recovering from major surgery, a nurse who was having trouble changing my complicated surgical dressing said to me, "I'm going to have to MacGyver this." I found the prospect unsettling, but I did tell her that she had just given me an example for a book I was planning to write. More recently, Bill Maher demonstrated on the New Rules segment of his comedy show that the term can also have negative overtones, "Obamacare? It should have been called MacGyver care" (*Real Time*).

MacGyver is certainly not the first such character in SF. Jules Verne's Cyrus Smith, identified as "the engineer," one of the castaways in *The Mysterious Island*, has been called the original MacGyver. Here he is demonstrating a lens he made to start a fire: "And he showed him the apparatus that he had used as a magnifying glass. It was two glass crystals that he had removed from his watch and the reporter's. After filling them with water and sealing their edges by means of a little clay, he had thus manufactured a real lens which, concentrating the sun's rays on some very dry moss, produced combustion" (84).

Of course, Smith and his fellow castaways have far more to do than usually falls to MacGyver's lot. Jules Verne says, "If, profiting from their acquired experience, they had nothing to invent, they nonetheless had everything to make. Their iron and steel was still only in the mineral state, their pottery was in the clay state, their linen and clothes were in the state of textile materials" (115). Still, they have just the kind of confidence in science and technology that MacGyver brings to any task (though they seem a bit more imperialistic than he is). As Pencroff, the sailor, says, "If you wish, Mr. Smith, we'll make this island a Little America! We'll build towns, railroads, telegraphs, and one fine day when it is transformed and civilized we'll offer it to the government of the Union!" (99).

Metalaw (also Interstellar Law, Space Law and Universal Law)

The *Oxford English Dictionary* defines metalaw as "A hypothetical legal code based on the principles underlying existing legal codes and designed to be applicable in a context or place (originally, space or outer space) not currently subject to legal regulation. Also, more generally: any law that operates beyond and outside known laws; a second or higher order of law."

The first use of the word was by "the world's first space lawyer, the late Andrew G. Haley ... at the Seventh International Astronautical Conference in Rome" in September 1956 (Freitas). His experience with space law went back to his participation with the Aerojet rocket company in 1942.

In 1956, Haley stated that "The maxim of metalaw ... is 'Do unto others as they would have you do unto them'" (*OED*). This was Haley's attempt to move beyond human-centered legal codes and ethics and put the emphasis on alien life forms whose values and comfort zones might be completely different. He continued that argument in his very thorough book *Space Law and Government* (1963). In a chapter titled "Metalaw," he wrote, "The indefinite projection of a system of anthropocentric law beyond the planet Earth would be the most calamitous act man could perform in his dealings with the cosmos. To extend our existing systems of law ... under the guise of an 'international law' which is to apply to 'outer space and celestial bodies,' would be to spread our terrestrial conflicts and intolerances wide and far through a universe that potentially offers tremendous vistas of a new age for man" (394).

Robert A. Heinlein used the word in his 1958 novel *Have Space Suit—Will Travel*, a narrative which contains a **bug-eyed monster** as a sympathetic character. In discussing the employment prospects of the main character, the Secretary General of the Federated Free Nations says, "A man skilled in space law and metalaw would be in a strong position" (247).

According to the *Great Soviet Encyclopedia* (1973), "The term 'space law' ... has become firmly established in official Soviet diplomatic documents and in the scientific writing of most of the socialist countries" (vol. 14, 759). The Soviets were clearly using the term in a somewhat limited sense, just as Heinlein seems to do. After all, diplomatic documents in the 1960s and 1970s did not deal with galactic issues.

The four different terms I have listed demonstrate either that there is some confusion about a single idea or that there are multiple ideas competing for room in this discussion. The term "metalaw" suggests that there may be a law underlying all other laws (or overarching them or informing them) and further that it would be possible to discover and make use of such a master pattern.

One of the clearest expressions of this notion comes from Clifford D. Simak's 1963 short story "New Folks' Home." This is not surprising. Simak's best work was set in the rural America of his youth, where the pioneer virtues of independence and neighborliness were tested against alien ideas and the aliens themselves. Again and again, as he does in *Way Station*, "The Thing in the Stone," "New Folks' Home," and many others, Simak tells the story of a human outsider who, with his feet firmly set on the rough beauty of Wisconsin farm country, meets and makes common cause with creatures from the stars. Some simple and good things, Simak implies, are truly universal. And by the time he's finished, he makes those things seem not only simple and good but also profound. He writes in "New Folks' Home," "Law could be approached in many ways.... As pure philosophy, as political theory, as a history of moral ideas, as a social system, or as a set of rules. But however it was viewed, however studied ... it had one basic function, the providing of a framework that would solve all social conflict" (97).

In Simak's story, a retired judge and legal scholar, who thought there was nothing left for him but an old folks' home, is given a house that restores his health and provides for all his physical needs. In exchange (or from his perspective, as an additional advantage), he will join an interstellar legal team as the only representative from Earth, indeed (and as usual with Simak), as the only person on Earth who knows there is such a thing. His first case is a "Summary & Transcript," a "Referral for Review Under Universal Law" (95). The argument that Simak makes is that since "there was throughout the universe a common chemistry" and therefore very likely "a common biochemistry as well" (97), "then did it not seem likely ... for there to exist a concept that would point toward common justice?" (98).

Simak does not suggest that those other intelligent beings are similar to humans. They "could not be expected to be men, or even close to men" (97). Nor does he argue that the goal of finding common justice would be achieved quickly, "Not just yet, perhaps. But ten thousand years from now. Or a million years from now." But Simak's retired judge believes very much in what he is about to do, "This was a thing he'd taught and preached for years—the hope that in some future time the law might represent some great and final truth" (98). It would be hard to suggest a more glorious vision of metalaw than that, even though Simak's words for it are different. He has already employed Universal Law. And the story's last sentence is, "He sat down and entered upon the practice of interstellar law" (98).

In 1970, Dr. Ernest Fasan in a book titled *Relations with Alien Intelligences: The Scientific Principles of Metalaw*, proposed a number of principles, based ultimately on the work of Immanuel Kant and suggesting that alien species could work together to achieve rational goals. Perhaps equally important (if not more so) was a thought experiment that began in 1966 and is still underway. I am referring, of course, to *Star Trek* in its many forms and to the cornerstone of metalaw (or as the *Star Trek* franchise calls it, perhaps with a less transcendent meaning, interstellar law) within the Federation, a doctrine of noninterference and nonintervention called the Prime Directive.

The idea behind it is that all societies are entitled to develop in their own ways without outside interference. Or in the words of *Star Trek Into Darkness*, "The Prime Directive clearly states there can be no interference with the internal development of alien civilizations." And the Prime Directive is every bit as inflexible as it sounds. In the film I just quoted, Captain Kirk is demoted for preventing an alien race from being exterminated by a volcanic eruption. Admittedly, in the process the technologically primitive aliens became aware of the existence of a starship, but surely a bit of inconvenient knowledge which will rapidly become folklore is a reasonable trade for survival.

Some earlier statements are, if possible, even harsher. In the original series episode "Bread and Circuses" (1968), Kirk, Spock and McCoy are being threatened by a "Roman" leader named Claudius (in one of *Star Trek*'s more ridiculous examples of "parallel" development). Kirk suggests that he could "bring down a hundred" of his crew "armed with phasers," and Claudius, who has inside information, replies, "You could probably defeat the combined armies of our entire empire, and violate your oath regarding noninterference with other societies. I believe you all swear you'll die before you'd violate that directive."

Of course, as with other thought experiments in long-running fictions (**Asimov's Three Laws of Robotics**, for example), the Prime Directive has more to do with creating plot difficulties than it does with smoothing out interstellar relations. One indication of this is the sheer complexity of what supposedly started out as Starfleet General Order 1. In the "Infinite Regress" episode of *Star Trek: Voyager*, Naomi Wildman (the precocious child for this series) is offering to recite lists of things from memory for the character Seven of Nine. She asks, "How about the suborders of the Prime Directive? I know all forty-seven of them." And just as importantly, we get to watch a range of Starfleet captains violate the Prime Directive and to consider the reasons they do so.

Captain Kirk, for example, may be the most ethnocentric of the bunch (though Kate Janeway has a long list of her own reasons for violating Starfleet's main rule). He has trouble accepting any society that is not dedicated to science, progress, and democracy. In the 1967 episode "The Apple," Captain Kirk squares off against a computer that runs a planet with, as Spock says, "a rudimentary intelligence." The inhabitants of the planet deliver to the machine the sources of energy it needs, and in exchange, they receive food, shelter and eternal life, though without sex, reproduction, higher thought and other complexities. At the end of the episode, McCoy justifies Kirk's destruction of a society that Spock compares to Eden by saying, "We put those people back on a normal course of social evolution. I see nothing wrong in that."

Of course, as Spock's sly suggestion that Kirk was the serpent in that garden of Eden makes clear, the series did not let anyone's values go entirely unchallenged, and that too was one of the reasons for introducing the Prime Directive into so many stories. As Spock points out earlier in the episode, "Doctor, you insist on applying human standards to non-human cultures. I remind you that humans are only a tiny minority in this galaxy."

No matter how rigid the Prime Directive may seem, there is something worse. In Arthur C. Clarke's novel *The Songs of Distant Earth*, the last ship from Earth is fleeing a dying planet, carrying one "million [people] who are ... sleeping," and heading for a new home called "Sagan Two" (170–171). Their criteria for selecting a new planet are remarkably restrictive and, in fact, are likely to reduce their ultimate possibility of success. The people of Earth had started with "the concept of Metalaw.... Was it possible to develop legal and moral codes applicable to *all* intelligent creatures"? (99). The conclusion was that "according to the principles of Metalaw, oxygen-bearing planets were placed out of bounds" (99). The argument was that "the presence of more than a few percent oxygen in a planet's atmosphere is definite proof that life exists there." That signals plant life or something similar, which may result in animal life, and that, in turn, may perhaps produce intelligent life.

Not only was it illegal to interfere with or influence a more primitive alien civilization, it was also forbidden to interfere with the process that might possibly result in such a civilization in the far future.

While it may seem that such extreme restrictions are unreasonable, there is a case to be made that if the goal is the preservation of alien civilizations and planets, the more restrictive the rules are, the better. In Rudy Rucker and Bruce Sterling's 2016 story "Totem Poles," extremely powerful aliens have come to Earth to help—at least that's how it looks from the results of their concerted effort to fix all of the planet's ecological problems. The humans who fight back against them (in spite of the good they are doing) have discovered that "The living saucers ... had a weakness. They carried within them some prime directive about intelligent life, some ethic that manifested itself as a tenderness towards human beings. The saucers were unwilling or even unable to harm people." The aliens were not, however, unable to interfere. The end result was "Twelve poles of supernal light, needles of prismatic brilliance, radiating into the cosmos, dissolving the substance of our world. Bathing in its native glow, the

Earth became a silver, dodecahedral orb, a mysterious cosmic traveler." I suspect most of the inhabitants of the Earth would have been in favor of a stronger and more detailed prime directive in that case.

There are, of course, certain omissions even from the most sweeping applications of metalaw. Both Simak and Clarke begin with the assumption that breathing oxygen is the only road to sentience or perhaps to life. Even on Earth there are organisms that belie the latter part of that belief. There is another assumption inherent in the term itself. Is it likely that anything will be universal in the far reaches of the galaxy? And even if law is universal at some stages of the development of sentient creatures, is it logical to think that no society will ever grow beyond it?

In the *Doctor Who* episode "A Good Man Goes to War," Madame Kovarian responds to the Doctor's expression of anger with, "The anger of a good man is not a problem. Good men have too many rules." Doctor Who says, "Good men don't need rules. Today is not the day to find out why I have so many." Is it not possible that somewhere in the far reaches of space and time we will find that statement to be true of an entire society and of laws?

Law professor Adam Benforado writes in *Unfair: The New Science of Criminal Injustice*, "The truth is that our descendents will be no less surprised by the ... systematic unfairness we tolerate today than we are by our ancestors' trials by ordeal. They will look back at our judges and juries and see biases that are just as obvious as the ones we now perceive in the bishops ... who presided centuries ago. They will look back at our criminal code and see laws as wrongheaded and illegitimate as the prohibition on heresy" (1).

It seems to me that there is a real possibility that an alien civilization might find "law" an extraordinarily primitive concept. They might say something rather like this, "Are we correct in believing that you require rules enforced by violence to prevent you from exterminating members of your own species? In other words, you achieve good behavior with menaces? Punish criminals with state-sponsored crimes? If so, please leave our solar system and do not come back." After all, in most of Simak's stories, humans and aliens interact successfully because of their inherent benevolence, not because of any laws they share, even when that benevolence (as in "A Death in the House") requires a certain level of self-sacrifice.

In 2015, Congress passed and President Obama signed the U.S. Commercial Space Launch Competitiveness Act. According to Planetary Resources, an organization with an extremely active interest in the legislation, "This law recognizes the right of U.S. citizens to own asteroid resources they obtain and encourages the commercial exploration and utilization of resources from asteroids" ("President Obama Signs Bill"). Eric Anderson, the Co-Founder and Co-Chairman of Planetary Resources, declared, "This is the single greatest recognition of property rights in history. This legislation establishes the same supportive framework that created the great economies of history, and will encourage the sustained development of space" ("President Obama Signs Bill"). In the words of Lulu Chang, "The new act paves the way for what some say could eventually become a multi-trillion dollar industry. Some estimates suggest that platinum filled asteroids could be worth as much as $50 billion."

While it could well be that almost anything which speeds up human expansion into space is ultimately good, there is a legal question here. Joe Palazzolo writes, "The Outer Space Treaty of 1967 ... says signatories should avoid 'harmful contamination' of the Moon and other celestial bodies, which are 'not subject to national appropriation.'" Still, he qualifies that by adding, "But the treaty is silent on private property rights."

Are we looking at a situation where shared international control (and de facto ownership of outer space) precludes individual property rights just as it prevents individual nations from staking territorial claims, or is it possible that individuals will be able to exploit the resources of space without their countries having control over any territory or they themselves owning anything except what they can carry away in their ships? In my second scenario there is an additional difficulty if prospectors go beyond asteroid mining to asteroid wrangling, where they seize asteroids, move them to new locations, and cut them up. Wouldn't that amount to claiming territory and not just property?

It seems as though it may be time for a new Space Treaty, one that recognizes that not all the ships in space will be launched by large national entities or, as time goes on, by large entities of any kind, by which I mean groups of inventors and entrepreneurs, like something from the early days of pulp SF. To give a rather surprising example, according to Jordan Rice from Astronomy.com, "A little-known company, Moon Express, announced that it received the 'okay' to go to the Moon as confirmed by the Federal Aviation Administration (FAA). The company currently has no plans to send humans back to the Moon, but it does want to send a robotic lander by the end of 2017."

Certainly, complex readjustments are necessary. In the words of Bignami and Sommariva from their 2016 book, *The Future of Human Space Exploration*, "If private companies are to seize the opportunities arising from space exploration, certainty over property rights and the uses of space resources are needed. This would involve extensive international cooperation" (3). With the new forms of competition now shaping up, that might prove a major understatement.

Multiplex Parenting

The term was coined in 2014 by César Palacios-González, John Harris, and Giuseppe Testa in their article for the *Journal of Medical Ethics* (Cook). They defined it as "a radical expansion of reproductive autonomy that allowed more than two persons to engage simultaneously in genetic parenting" (5). Sonia Suter, a legal expert from George Washington University, gave a more detailed definition. According to her, multiplex parenting occurs when "groups of more than two individuals (whether all male, all female, or a combination) procreate together, producing children who are the genetic progeny of them all" (Suter 2).

The process became possible in 2006 when Shinya Yamanaka discovered that cells from adults could be turned into stem cells: "By introducing only a few genes, he could reprogram mature cells to become pluripotent stem cells, i.e. immature cells that are able to develop into all types of cells in the body" ("The Nobel Prize"). For his discovery, Yamanaka shared a Nobel Prize in Physiology or Medicine in 2012.

By 2014 it became clear that with Yamanaka's discovery, scientists could produce sperm and egg cells outside the human body using cells from adults. That, in turn, made it possible to create embryos that have two, three, or thirty genetic parents. Almost any cell from any individual could be turned into a sperm cell or an egg cell. These would be combined to create an embryo. From the cells of that embryo, another set of sperm cells or egg cells could be produced. In a three-parent scenario, either a sperm cell or an egg cell derived from the embryo would be combined with either a sperm or an egg of the third parent. In a four-parent case, first two embryos would be created, using genetic material from all four people, two for each embryo. Then, a sperm cell would be made from the cells of one embryo and an egg cell from the cells of the other. Next, the sperm and egg cells would be combined to make an embryo that is related to all four people who donated the initial cells.

There will be, of course, questions of law. Could these four people qualify as the legal parents of this child? Today, experts say no. "The resulting children would therefore be 'orphaned,' since their genetic 'parents,' the embryos, would have died before the child was even born" (Suter 24). The four people involved would not be parents but grandparents, "The resulting embryo would be genetically related to all four prospective parents, who would technically be the child's genetic grandparents" (Palacios-González et al. 5).

The technology for multiplex parenting would make it possible "to create 20 or 30 generations of Petri dish humans in as little as ten years" (Cook). Such a process would make the legal issues even more complex, but it would also provide opportunities for genetic manipulation on a breathtaking scale. Given the abilities we have now (and more importantly soon will have) to manipulate genomes, the creating of thirty generations in ten years could make possible a transformation beyond what most SF writers have imagined. (See **Mutant**.) At the very least, it could open the door to what I call veneer babies, children who are smarter, stronger, longer-lived, better looking, more resistant to all diseases and immune to many, and so on but with a veneer of DNA from their (probably two) "parents" which would make them look like members of their immediate family.

On a more limited note (in many senses of that word), multiplex parenting would also make it feasible for only one person to reproduce without the aid of a partner—in effect, **solo parenting**. Both sperm and egg cells could be made from the cells of one adult. The resulting embryo could then be implanted into a woman or put into a **uterine replicator**.

For those who think that nothing like this will ever happen, there have already been successful experiments with mice, and "progress in mice has not revealed any fundamental hurdle that should impede success also with human cells" (Palacios-González et al. 2). More than that, there is a parallel human case which has already been approved and carried out in Britain. "Mitochondrial replacement therapy" is a medical procedure "whereby healthy mitochondria from an unrelated donor are added to an existing embryo" (John Harris). The procedure is recommended for potential parents with mitochondrial disorders which may subject their children to a variety of debilitating and even fatal diseases.

Science fiction has explored the possibilities of multiplex parenting in various ways. For example, William Tenn in his story "Venus and the Seven Sexes" has aliens that produce children using seven biological parents, "A chain" of parents representing seven sexes is carefully selected to "produce offspring of maximum variability. Then ... when the chain is established, each sex begins to secrete in its original germ with the full forty-nine chromosomes." Greg Egan in *Diaspora* imagines multiplex (and solo) parenting in a society of citizens who are essentially "conscious software." Their children are created by "non-sentient software" from "one parent, or two, or twenty" (5). SF writers have suggested that altering sexual roles and changing the process of childbearing and parenting (Ursula K. Le Guin in *The Left Hand of Darkness*, for example) could significantly impact the structure of society. It looks as though we will soon have the opportunity to see if those predictions are correct.

Mutant

The *OED* gives as its second definition, "Science Fiction. An individual imagined as having arisen by genetic mutation, *esp.* one with freakish or grossly abnormal anatomy, abilities, etc." A more concise definition would be: A being whose genetic mutation has gone terribly wrong or terrifyingly right. In effect, science fiction writers and futurists have explored both the dangers and the opportunities. And ultimately, the most important question they have raised is the one about the nature of new species of humans, what Philip K. Dick, among others, calls "homo superior" (*The Collected Stories* 44).

The word "mutant" goes back to Hugo de Vries and his theory of mutations in various works at the turn of the last century. In *Species and Varieties: Their Origin by Mutation* (1906), he says, "These were related as 'nieces' to the first observed mutants" (476). However, the science fiction definition, first identified by the *OED* in 1938, is at once more and less than that.

What passes for mutation in science fiction or in comics and films is very often something different, a teratogenic or congenital effect (or defect) that does not change the genes and is, therefore, not inheritable, or instead, a kind of transformation that is more symbolic than real but is no more explicable than Kafka's cockroach. Clearly, the word contains more than one idea, though they are deeply intertwined. One notion behind "mutant" is that a mutation can be enormously strange, terrible, or even awe-inspiring and that it can leap into being in one sudden event, a birth or a lab accident. Even though a single event that creates a group of mutations which work effectively together has long been held to be nearly impossible and the things that are supposed to have brought it about in fictional worlds are almost always laughable, there have been (and will continue to be) many such stories.

One reason the number and prominence of these stories is increasing is, as Joel Garreau argues, "In the middle of the 20th century, the powers of these superheroes were dreams. Today, we are entering a world in which such abilities are either yesterday's news or tomorrow's headlines" (21).

Many SF uses of the word "mutant" linked in some way to radioactivity, though it's true that similar notions existed long before the atom bomb. For instance, Oliver Wendell Holmes's novel *Elsie Venner* (1861) concerns a girl (she dies at eighteen) who has some of the powers (real and imagined) and in part the nature of a serpent because her mother was bitten by one during pregnancy. Elsie kills a governess with poison in some strange and unexplained way. While looking for Elsie, Bernard Langdon, the hero of the novel, finds himself, stunned and powerless, about to be struck by a rattlesnake. Instead, "He heard a light breathing close to his ear, and half turning, saw the face of Elsie Venner, looking motionless into the reptile's eyes, which had shrunk and faded under the stronger enchantment of her own" (141).

Whether Elsie's transformation is teratogenic or mutagenic is not clear, since she does not have a child, but the pattern of a human acquiring powers from the bite of a lower life form is familiar. Except that the bite of a poisonous snake takes the place of a radioactive spider (and that Peter Parker receives his powers directly), Elsie Venner as a snake girl and Peter Parker as Spiderman are very similar. The big difference is that Stan Lee "like most people in the Fifties, was fascinated by radiation" (Knowles 139), and so the spider's strange gift was connected to radiation. Exactly how a nearly adult human could be so transformed is never explained, but metaphors and symbols do not usually come with scientific explications attached, even when they are not in comic books about superheroes. In much the same way, a long list of monsters who had been transformed, awakened, or just severely annoyed by radiation filled the movie theatres of the 1950s as one manifestation of what has been called *fictions of nuclear disaster*.

Because most mutations of any kind, and nearly all of the other effects of exposure to radiation, are negative or deadly, mutation was one of the many fears attached to nuclear power and atomic bombs. And as usual, science fiction was not just about myths and metaphors, it was also about making accurate predictions and creating realistic examinations of the possible worlds that might result. "As early as 1942, Lester del Rey wrote a novella (published in *Astounding*) about a meltdown at a nuclear power plant." Called "Nerves," it "presupposes a world where atomic power is an accepted fact, where atomic-power plants are as commonplace as automobile factories" (Healey and McComas 47). That was "three years before the first bomb and twelve years before the world's first 'nuclear powered electricity generator began operation' in Obninsk, Russia" (Pilkington "Forbidden Planet" 51). (See **Atomics**.)

"Tomorrow's Children," Poul Anderson's first published story (*Astounding* 1947) describes the aftereffects of a sustained nuclear conflict. Civilization has been shattered. General Robinson, acting President of the United States, identifies a calf as a "mutant." He continues, "There are even a lot of human abnormal births.... In fact, that's just about our worst problem" (113). For humans, "about seventy-five per cent of all births are mutant, of which possibly two-thirds are viable and presumably fertile" (123).

There are many suggestions as to how to deal with the problem, most of them proposed by Robinson, including killing the mutant children, sterilizing them and their parents, creating a "Eugenics" program to ensure that humans breed with humans, or placing the humans in protected areas. Drummond, the story's main character, rejects them all as impracticable, and he and Robinson eventually conclude that the best thing they can do is follow Drummond's suggestion to "abandon class prejudice and race hate altogether, and work as individuals. We're all ... well, Earthlings, and subclassification is deadly. We all have to live together, and might as well make the best of it" (127).

In Lois McMaster Bujold's Vorkosigan series, the planet Barrayar survived a war during which their enemies used nuclear weapons. In *The Warrior's Apprentice* (1986), Miles Vorkosigan jokes about land he owns where "you can still see a faint glow in the sky, on a dark night, twenty kilometers off" (58). Not surprisingly, there is a great fear of mutation, with the less sophisticated Barrayarans treating anything visibly different as though it might be catching, and even practicing infanticide. Miles, who has a number of obvious problems caused by a poison gas attack on his mother while she was pregnant with him, is forever explaining, "It's teratogenic, not genetic, I'm not a mutant" ("The Mountains of Mourning" 278). Even so, when Ungari, Miles's immediate military superior, who definitely knows the truth, is pushed to the breaking point, he says, "Vorkosigan! You mutinous little mutant" (*The Vor Game* 547).

Bujold's Vorkosigan series is consistently realistic in its presentation of the effects of radiation. There are no giant monsters or powerful superheroes who result from exposure. The stories where such things happen are either making individual monsters to stand in for the general dread of nuclear power and its accompanying radiation or they are suggesting that life might find in mutation a way of transcending what can seem to be inevitable death.

Not surprisingly, such stories frequently substitute the fear of mutants for the fear of radiation, making human mutations, in many cases, the same sort of stand-ins for the general dread of nuclear power as other monsters that supposedly result from or are somehow connected to radiation. Bujold's Barrayar books show that, realistically, even Miles Vorkosigan or a baby with a cleft palate may be taken for such a creature. In his own poetic epigraph to *New Maps of Hell: A Survey of Science Fiction* (1960), Kingsley Amis expresses the fear: "But climates and geographies soon change,/ Spawning mutations none can quell/ With silver sword or necromancer's ring,/ Worse than their sires, of wider range/ And much more durable" (xii).

This fear behind the word "mutant," is also the dread of the other, and in this case the fear of a different creature who may destroy humans, because either the mutants are monstrously dangerous in some way or they are superior in evolutionary terms and will inevitably cause the extinction of their inferiors. Not everyone felt this way about mutants, however. As Philip K. Dick points out, "In the early Fifties much American science fiction dealt with human mutants and their glorious super-powers and super-faculties by which they would presently lead mankind to a higher state of existence, a sort of Promised Land. John W. Campbell, Jr., editor of *Analog*, demanded that the stories he bought deal with such wonderful mutants" (*The Collected Stories* 411).

This is an aggressive suggestion of the transcendence of dangers, the elimination of terrors, not only for the mutants themselves but also for the rest of humanity. Radiation, which had seemed to be nothing but deadly, becomes a bridge to the superhuman, a way when no other seemed possible.

Philip K. Dick, of course, was commenting on one of his own stories which was the exact opposite of Campbell's requirements. In "The Golden Man" (originally published as "The God Who Runs" and filmed as *Next* in a much different form with Nicolas Cage), Dick describes what is very nearly a parody of Campbell's messiahs. In a world where mutants caused by a nuclear war are hunted down and killed, there is finally one who can survive. "He doesn't talk," but "he's incredibly beautiful. A god come down to earth" (40).

The mutant doesn't have language or human intelligence, but he has two powers. He is irresistible to women, and he can see ten minutes into the future. That's enough to let him survive. If his genes are dominant and not recessive, it's enough for his kind to prevail over humans. Of course, Dick was making the point that evolutionary survival is not necessarily about what we think of as superior. Still, his demonstration of the possible replacement of one species by another is instructive even if his version is highly unlikely.

Lester del Rey had explored that same question more plausibly in 1939 in a story first published in *Astounding*. "Day Is Done" recounts the death of the last Neanderthal in the presence of (but not quite at the hands of) Cro-Magnons. Legoda, one of the new men who is sympathetic to the Neanderthalers, says, "Your people die too easily ... no sooner do we find them and try to help them than they cease hunting and become beggars. And then they lose interest in life, sicken and die" (74). Del Rey makes it clear that the Cro-Magnons have marginalized their competitors—taken their hunting grounds, shown them up as inferior toolmakers and hunters, and made it clear that they can survive only on what the new men let them have in charity or through carelessness.

The notion of one group's being discouraged by the superiority of another is common in science fiction. In "Asylum" (first published in 1942), A. E. Van Vogt describes Kluggs, who are very near the bottom of the Galactic intelligence scale and are given the lowest of tasks (such as overseeing Earth). Leigh says, "'Klugg, Klugg, Klugg! So you realize now that the Dreeghs had you down pat.... 'Because if you're I. Q. 243, the Dreeghs were 400' ... She was as white as a sheet. The astounded realization came that he had struck, not only the emotional Achilles heel of this strange and terrible young woman, but the very vital roots of her mental existence" (633).

Far beyond the Dreeghs are Great Galactics with I. Q.s of 1,200 (644). In a later story (1965), Van Vogt shows a reaction to *them*, "He had seen one of the Great Ones, and it was not a memory that it was good for a lesser being to have. Somehow it hurt to be so much less" ("Research Alpha" 209).

In Poul Anderson's "Time Patrol" (1955, the first in that series of time travel stories), the very confident Dard Kelm is introducing students to the Academy, "For the first time, his casual, half-humorous air dropped, and he stood there as a man in the presence of the unknowable. He spoke quietly: 'The Danellians are part of the future—our future, more than a million years ahead of me. Man has evolved into something ... impossible to describe. You'll probably never meet a Danellian. If you ever should, it will be ... rather a shock. They aren't malignant—nor benevolent—they are as far beyond anything we can know or feel as we are beyond those insectivores who are going to be our ancestors. It isn't good to meet that sort of thing face to face'" (Anderson *The Time Patrol* 7).

On a lighter note, in *Gene Roddenberry's Andromeda*, Dylan Hunt says, "Ah, Nietzscheans! It must be nice to be genetically perfect." (See **Kludge**.) And Trance, herself a powerful alien, replies, "I like your flaws, Dylan." His response, which lacks any enthusiasm, is, "Oh, thank you!" ("A Symmetry of Imperfection").

In one of the neatest blendings of the fears of nuclear weapons and superhuman successors, the 1957 film *The Abominable Snowman* suggests that a superior race will replace humans after a nuclear holocaust. The lama of a Buddhist monastery in the Himalayas tells Dr. John Rollason, the film's main character, "Man is near to forfeiting his right to lead the world. He faces destruction by his own hand." Rollason initially speculates, "There's a warning in the creature. It's strong, intelligent. It may have powers we haven't even developed. It might have inherited the earth, but something went wrong. And here it is the last vestige of a species hiding away where nothing else will live."

Tom Friend, the cynical leader of the expedition, points to the same moral as the lama, "Look, if they drop the H-bomb, your descendants too may wind up in the ice." At last Rollason understands, "Suppose they're not just a pitiable remnant waiting to die out. They're waiting, yes, but waiting for us to go." The film leads the audience to conclude that if these long-lived giants who have both sadness and wisdom were to replace humans, the world would be the better for it.

In the face of such fears and discouragements, it may seem difficult to explain the continuing fascination with superior mutants. It is more than just a particularly compelling form of horror story. In explaining why people wanted positive stories about powerful mutants, Philip K. Dick says, "I think these people secretly imagined they were themselves early manifestations of these kindly, wise, super-intelligent Übermenschen" (412). Given the fact that science fiction fans have frequently referred to themselves as slans, after the superhuman mutants in Van Vogt's novel of that name, Dick seems to be correct. But it is also the desire to transcend the danger of radioactivity and to explore the possibility of something beyond humans as they now exist.

There has never been a shortage of superhumans in science fiction. Nor have all such beings been the result of natural evolutionary forces or the unnatural mutations induced by radiation. From Olaf Stapledon's *Odd John* to A. E. Van Vogt's *Slan*, from Harry Kutner's *Mutant* to Gordon R. Dickson's Dorsai series, from Nancy Kress's *Beggars in Spain* to Lois McMaster Bujold's *Cetaganda*, from the X-Men franchise to the *Heroes* series, there has been a parade of possibilities, a series of suggestions of what human beings might be with just a little transformation. There were even two television series about mutants created by geneticists, albeit not kind and responsible ones, *Mutant X* and James Cameron's *Dark Angel*.

As long ago as 1921 in *As Far as Thought Can Reach*, George Bernard Shaw was looking forward to humans who had evolved themselves "to the vortex freed from matter, to the whirlpool in pure intelligence that, when the world began, was a whirlpool in pure force" (261). They had gone beyond "flesh and blood," saying, "It imprisons us on this petty planet and forbids us to range through the stars" (255).

However, for some time it seemed that such stories, unless they could be explained by the strange impact of radiation (or Shaw's version of creative evolution), were impossible. Writing in 1991, Roberta Rogow expressed what was then the conventional wisdom, "In the last eight thousand years or so, culture has taken over the function of physical evolutions, and those mutations that have actually occurred may be so lethal as to eventually die out entirely" (218). In other words, we have reached the end of change and transformation. But science fiction and its preoccupations were closer to the truth. Researchers using modern genetic techniques on human remains from 8,000 to 2,000 years old, have come to very different conclusions: "The scientists noted thousands of distinct places where there were DNA changes that didn't seem just random ... but a sign that the human species was actually adapting to some aspect of its environment" (Whoriskey). There were "12 places in the DNA where the signs of selection were clustered. One of those clusters related to the digestion of milk—a confirmation of previous findings. Another related to how humans digest fatty acids.... Others were related to skin color, which got lighter as humans moved northward from Africa" (Whoriskey).

While these were not harbingers of homo superior, they indicated that culture might drive evolutionary change as well as slow it. As Enriquez and Gullens put it, "As natural selection gave way to human selection, we adapted so quickly to our changing environment that a full 7 percent of our genes underwent rapid evolution in the last 5,000 years" (258).

Now, however, those other forms of mutation, those other mutants, the kind we choose to create, are nearly within our reach. We will not, perhaps, produce vortices freed from matter, but the time when there is one and only one human species may be rapidly nearing its end. There was a point in the past when there were multiple hominins. "At least three of the 'other species' were alive when our ancestors emerged from Africa 50,000 to 80,000 years ago: Denisovans, Hobbits, and Neanderthals" (Enriquez and Gullans 228). Many modern humans have DNA from Neanderthals and Denisovans. The day may be coming when transhumans will talk about having DNA from *Homo sapiens sapiens*. Even as I write this, "We can design, build, and transfer new genes into humans within months.... Millions of years' of evolution is being reformulated, redirected by humanity in just a few years" (Enriquez and Gullans 259).

The speed will increase, the possibilities will multiply, and there is even a vanishingly small chance that we might reach Bruce Sterling's unsettling future when "not only is humanity extinct but, strictly speaking, pretty much everyone alive today should be classified as a unique, postnatural, one-of-a-kind species" (442). Or we might be looking at the future outlined by Michael Goldblatt, then head of DARPA, the United States' Defense Advanced Research Projects Agency (Garreau 19). He described it as "science action, not science fiction" and said, "Imagine if soldiers could communicate by thought alone.... Imagine the threat of biological attack being inconsequential. And contemplate, for a moment, a world in which learning is as easy as eating, and the replacement of damaged body parts as convenient as a fast-food drive-through" (cited in Garreau 22–23).

It may be, instead, that we will not view the future through a military lens but that there will be benevolent superhumans and paradises beyond imagination. It is very likely that the future holds life forms beyond our dreams because scientists have already built a new form of DNA with "two new base pairs" that "turned a four-letter code of life into a six-letter code" (Enriquez and Gullans 255). As Juan Enriquez says,

"This Lifecode stuff ... is the single greatest superpower humans have ever had" (Enriquez). Soon, it will indeed be true that a suite of mutations can leap into being in one sudden event in a lab, that the resulting mutant can be a new species, better than the one that came before, and perhaps it will "be nice to be genetically perfect."

Nanotechnology (also Nanotech)

The manipulation of matter at the atomic level (or 1 to 100 nanometers), most commonly visualized as truly tiny machines (or alternatively chemical, biological or other processes) maneuvering individual atoms into predetermined patterns. The *OED* Online gives N. Taniguchi credit for the first use of "nanotechnology" in 1974. It was popularized (some might even say propagandized) in 1986 by K. Eric Drexler in *The Engines of Creation*, but as so often with the terms in this book, the idea behind "nanotechnology" did not begin with the word itself. German physicist "Arthur von Hippel ... had such ideas in the 1930s.... 'Nature designs everything from atoms,' he said, 'Hence we should be able to design any kind of material with foresight'" (Regis *Nano* 71–72).

Von Hippel had a long and distinguished career. He was, among many other things, a professor at MIT, the coinventor of radar, coiner of the term "molecular engineering" (regarded by some as more precise than "nanotechnology"), and author of the 1959 book *Molecular Science and Molecular Engineering*. In 1962, he wrote in *Science*, "Molecular designing allows us to *realize* [emphasis his] Jules Verne's fantasies. The answer is therefore not any longer what we can do, but what we want to do" (108). I pause here to point out that this sort of connection between science fiction and scientists is not so much common as commonplace.

It is, however, Richard Feynman who is usually called the "father of nanotechnology." He was a flamboyant personality, a Nobel Prize winning physicist and an extremely successful futurist (in large part because his colleagues tended to believe in his predictions and to make them happen). James Gleick calls him, "The most brilliant, iconoclastic, and influential physicist of modern times" (8). Feynman said in 1959, "The principles of physics, as far as I can see, do not speak against the possibility of maneuvering things atom by atom" (137). He also declared, "All of the information that man has carefully accumulated in all the books in the world can be written ... in a cube of material one two-hundredths of an inch wide—which is the barest piece of dust that can be made out by the human eye" (123). He all but demanded that physicists "make the electron microscope 100 times better" (124), and suggested that if computers could be miniaturized and had "millions of times as many elements, they could make judgments" (127).

Feynman was addressing the annual meeting of the American Physical Society (held in 1959 at Caltech), and he discussed information storage and manufacturing at the atomic level, using DNA as an existing example. He said, "I want to build a billion tiny factories, models of each other, which are manufacturing simultaneously" (134). He also offered two $1000 prizes, "one for the first microscope-readable book page shrunk 25,000 times in each direction, and one for the first operating electric motor no larger than a 1/64th-inch cube" (Gleick 355). Feynman didn't have to pay the first prize until 1985; the second challenge, however, was successfully met in 1960, and Feynman "who had neglected to make any arrangements for funding ... sent ... a personal check" (356).

If nanotech hasn't yet reached the full glory of Feynman's vision, it has moved from the realm of inspired speculation to hard (if small) fact, and as early as 1990 the IBM laboratories in Zurich used their scanning, tunneling microscope to spell out (What else?) IBM with 35 xenon atoms (Drexler 3–4). In the twenty-first century, nanotech has gone mainstream with massive investments from governments and an ever-proliferating list of applications and research institutes, everything from the treatment of cancer to the salvation of the environment. Books on the subject are piling up, with titles such as *Nanotechnology for Dummies*, *Understanding the Nanotechnology Revolution* and *Nanotechnology and Homeland Security: New Weapons for New Wars*.

Nanotech has also, of course, become part of the furniture in many SF stories, but some writers have had difficulty visualizing the "size" of the change inherent in this technology. Imagine a world, not of scarcity as ours has always been but of plenty, where everyone could have almost everything. Imagine a world where even the poorest human would live better than American millionaires do now. In a limited number of ways, such a thing has already happened as technologies advanced. Most inhabitants of industrialized countries have transportation, entertainment and information available to them which would be the envy of kings and emperors from earlier times. Few millionaires in the 1940s had film libraries to match those in or immediately available to most middle-class homes today.

Nanotech (if it turns out to be technically feasible but not prohibitively high in energy cost) would similarly transform the lives of almost all humans. Anything (including art masterpieces and DNA-specific human organs) could be manufactured from the atoms up. Raw materials could be mined from the sea, from garbage dumps or from asteroids. Space explorations and space colonies would finally be cheap, justifying at last the enthusiastic visions of Eric Drexler and Gerard K. O'Neill (McCray 10). Nearly unlimited goods and services (most manufactured or performed by robots and programmable machines of various sizes) would become part of life for nearly everyone.

To take a small example, the January 2000 issue of *Discover Magazine* predicted nanotech clothing that would be manufactured in nano-mills the size of photocopiers (Marston 48). The clothing would be self-repairing and self-cleaning.

The "Empathy for the Devil" episode of *Extant* (2015) presented just such a mechanism for making tee shirts and dresses. And while that was fictional, desktop 3D printers, primitive harbingers of the machines to come, are already for sale on Amazon and at Staples. The capabilities of the comparatively new technology are increasing with surprising speed. To give a single example, "GE engineers working on the future of aircraft manufacturing recently ... made a simple 3D-printed mini jet engine that roared at 33,000 rotations per minute" (Keller).

And to get back to nanotech itself, the 2016 Nobel Prize in Chemistry went to Jean-Pierre Sauvage, Sir J. Fraser Stoddart, and Bernard L. Feringa "for their design and production of molecular machines." The Press Release from the Royal Swedish Academy of Sciences continued, "They have developed molecules with controllable movements, which can perform a task when energy is added." They calculated that "the molecular motor is at the same stage as the electric motor was in the 1830s, when scientists displayed various spinning cranks and wheels, unaware that they would lead to washing machines, fans and food processors" ("Press Release"). Of course, the speed of technological change is far greater now than it was in the nineteenth century, and it is constantly accelerating.

One of the many end results of the coming of nanotech will be that money (as people keep saying in *Star Trek*) will cease to have meaning, and working for a living, as we presently understand it, will no longer be necessary—or possible. The people on technology's cutting edge can already see that new world coming, "Martin Ford cites a recent jobs summit he attended with about fifty tech-company CEOs. 'Here in Silicon Valley, there's a remarkable consensus about this. Every single person agreed we're on the leading edge of a disruption, and we're going to have to move to a guaranteed basic income. There was overwhelming support for that'" (iv).

As almost every episode of every series in the *Star Trek* franchise makes clear, however, it is tremendously difficult for writers to create stories that do not involve the scarcity of property as a given. One of the better attempts to come to terms with this enormous disparity in mindsets (and one of the better *STNG* episodes) is "The Neutral Zone," where the crew of the *Enterprise* has to explain the shape of the future to some twentieth-century humans whom Beverly Crusher, the ship's chief medical officer, has revived from cryogenic sleep. (See **Cryonics**.) Captain Picard says, speaking to a millionaire whose money has vanished over time, "This is the 24th century. Material needs no longer exist." And the former millionaire replies, "Then what's the challenge?" Picard's answer is, "The challenge, Mr. Offenhouse, is to improve yourself, to enrich yourself. Enjoy it."

It's either that or the holodeck, but most *Star Trek* writers (and the Ferengi) seem never to have heard of a replicator, which is *Star Trek*'s equivalent of nanotech (though nanotech does come in under its own name now and again). Episode after episode sends the *Enterprise* (in its many incarnations) searching desperately for some substance that they could easily replicate—or build from the atoms up. But admittedly, this is also the sticking point for many people who are not *Star Trek* characters. The possession of things has meant survival, power and prestige for so long that it's hard suddenly to dismiss them as what they may be about to become—arrangements of atoms that we will control as effortlessly and treat as casually as the digital download of this book.

I am not suggesting that there are no dangers attached to nanotech. A system that depends on **robot**s of various sizes and looks toward the Singularity (the point when computers become smarter than humans) as "a consummation devoutly to be wished" has room for most of the disasters that fill the nightmares of neophobes. And even if nothing truly terrible happens, the road to a better world is almost always paved with disappointments. No society has ever gone through a dislocation as large as this one promises to be, and predicting what it will be like is exceedingly difficult even for our best science fiction writers and futurists.

But the end result promises to be extraordinary. Imagine human lives set free from hard work and from worry about survival, human beings who are free to pursue old values and new opportunities. Imagine an end to most of the problems that have made human lives miserable—poverty, drudgery, war and disease. And the ending of disease may be the first step in the elimination of old age and a remarkable increase in life expectancy. This is likely to be a future worth the struggle.

Neural Lace

A neural interface between the brain and electronic or chemical systems that exists as an implanted "grid of wires only a few millimeters across" (Powell). There have been other types of neural connections before, but they have been larger and required surgery. For instance, it was reported in 2015 that "Dr. Paul Nuyujukian, a neuroengineer and physician from Stanford University" had been successfully working on "neural prostheses." With that technology, "a baby-aspirin-sized microarray chip is directly implanted into the brain, and neural signals associated with intent can be decoded by sophisticated algorithms in real time and used to control mouse cursors" (Fan).

Impressive as that is, neural lace is a "paradigm-shifting approach" (Hong et al.). It was "developed by researchers in Charles Lieber's lab at Harvard University and the National Center for Nanoscience and Technology in Beijing" (Kurzweil). "The mesh is ... tightly rolled up, allowing it to be sucked up into a syringe via a thin glass needle" (Kurzweil). Once it is injected into the brain, the mesh opens up to 30 times its compressed size, and brain cells grow around it without damage. "Research study leader Charles Lieber" says, "They are one million times more flexible than any state-of-the-art flexible electronics and have subcellular feature sizes. They're what I call 'neuro-philic'—they actually like to interact with neurons" (Kurzweil).

There is a whiff of very large possibilities here, including enhanced intelligence, access to virtual worlds, nearly infinite awareness during movement through the real world (or universe), and a new means of monitoring and controlling personal health and longevity.

At this point, though, there is not much more than a proof of concept, and the brains in question belong to mice. However, this is the first step in the actualization of a science fiction idea and accompanying term created by Iain M. Banks in his "Culture" series of high powered, semi-optimistic space operas. Mary-Ann Russon writes, "If the neural lace is able to completely integrate with the human brain, this would enable doctors to treat all sorts of neurodegenerative diseases that are currently difficult to cure." She then goes on to suggest a slightly larger possibility, "There's also the sci-fi element of cyborgs— humans that have electronic parts in them—becoming a reality, as well as the concept of humans having brains that can instantly tap into infinite knowledge."

Nor is she the only one to have made such a suggestion. Elon Musk has said that neural lace may be the solution to a very large problem. He says, "If you assume any rate of advancement in AI, we will be left behind by a lot.... We would be so far below them in intelligence that it would be like a pet ... like a house cat." The answer is to increase the capabilities of human beings. Musk argues, "The solution that seems maybe the best one is to have an AI layer.... [A] ... digital layer could work symbiotically with the rest.... It'd be some sort of direct cortical interface ... a neural lace.... If we can figure out how to establish a high bandwidth neural interface with your digital self effectively, then you're no longer a house cat."

It may seem a big leap from experiments with mice to re-establishing humanity's place in the face of a challenge from ultra-intelligent AIs, but "Harvard's Office of Technology Development has filed for a provisional patent on the technology and is actively seeking commercialization opportunities" (Kurzweil). And when Elon Musk was asked if he would be interested in "exploring the possibility" of creating the company that would make the neural lace a reality, he said, "Somebody's gotta do it.... If somebody doesn't do it, then I think I should probably do it."

The first reference to "neural lace" itself seems to be in *Excession* by Iain M. Banks (1996). It is no surprise that most articles on this new development labeled it with the term he created and popularized. Gizmodo's headline was, "Scientists Just Invented the Neural Lace." The article begins, "In the Culture novels by Iain M. Banks, futuristic post-humans install devices on their brains called a 'neural lace.' A mesh that grows with your brain, it's essentially a wireless brain-computer interface" (Newitz).

Banks' explorations of the subject have covered a wide range of possibilities. In *Look to Windward*, we get a sense of just how much a part of people's lives neural lace has become, "After one of those vacant expressions you got used to when people were consulting a neural lace or other implanted device—she smiled.... 'Oh,' he said, surprised. 'Do you have a neural implant?' ... Being fitted with your first implant was about as close as some bits of the Culture got to a formal adult initiation rite."

There's a truly extreme case of memory storage in Banks' *The Hydrogen Sonata*: "QiRia said ... 'In some ways my brain is as it's always been, just stabilised. Been like that for millennia. Though it does have a modified neural lace within it. Heavily modified; no comms. What I do have is extra storage. Not processing; storage. The two are sometimes confused.... It's in me.... Throughout me. Vast amount of storage room in the human body, once you can encode in the appropriate bases and emplace a nano-wire read-out system through the helices" (172–173).

There are also examples of humans rising to something approaching the level of AIs or as Banks calls them "Minds." In *Surface Detail*, there is a "neural-laced-brain running at as near to AI-speed as beyond-humanly possible" (449). In *Matter*, Banks writes, "She ... clicked into her neural lace and through it into the vast, bludgeoningly vivid meta-existence that was ... the Culture's dataverse. It was a given that this perceptual frenzy was as close as a human, or anything like a human, could get to knowing what it was like to be a Mind" (94).

But perhaps the most extraordinary capacity of the neural interface is showcased in this dialogue from *Surface Detail*, "'A full back-up-capable neural lace grows with the brain it's part of, it beds in over the years, gets very adept at mirroring every detail of the mind it interpenetrates and coexists with. That's what you pretty much must have had. Plus it had an entanglement facility built into it, obviously.'

"She glared at Sensia. 'So I'm ... complete? A perfect copy?'

"'Impossible to be absolutely sure, but I strongly suspect so. There is almost certainly less of a difference between the you that died and the you that you are now than there would be between your selves at one end of a night's sleep and the other'" (78).

Questions of identity apart, the complete and accurate duplication of an individual is a major accomplishment for any piece of technology. (For some indication of the difficulties see **Teleportation**.) Plus, as usual, science fiction has furnished thought experiments that provided ideas and terms for innovations yet to come.

I do not mean to suggest, of course, that "neural lace" is the first reference in science fiction to a direct connection between the human brain and computers or other technology. Neural lace itself does not appear in the *OED*, Prucher's *Brave New Words*, or *The Encyclopedia of Science Fiction*. The *OED* indicates Michael Bishop's *And Strange at Ecbatan* in 1976 as the first appearance of the SF meaning of "neural": "The ones who have roles always require surgical adaptation, electrode implanting, cybernetic neural grafting." The *OED* definition is "Connected directly to the nervous system; relating to or designating an interface between a device and the nervous system."

Jeff Prucher has a reference antedating that from *Grandpa* by James H. Schmitz in 1955: "The countless neural extensions that connected it now with the raft came free" (129).

Neither the *OED* nor Prucher include an example from William H. Gibson's cyberpunk bible *Neuromancer*, from 1984: "What would you say if I told you we could correct your neural damage, Case?" (29). Not only does the word turn up, but the surgical procedure seems to have proceeded along familiar lines for those who have been reading about the real beginnings of neural lace, "The nerve stuff.... Lot of injections. They didn't have to open anything up for the main show" (32). It's possible certainly to read "neural" there as having to do simply with the brain and not with anything connected to it, but in the context of a hacker who's being refurbished to do a very big hack (after his employers had punished him for stealing by fixing it so he could feel "his talent burning out micron by micron" [6]), it seems unlikely. (See **Cyberspace**.)

Dictionaries (even extraordinary historical dictionaries like the *OED* and Prucher's SF offshoot) have other goals than mine. It's words they want more than ideas, and rightly so for them.

Finally, for my examples at least, in 1986, Gregory Benford and David Brin use "a direct neural link" in *Heart of the Comet* (7). It's clear almost immediately, though, that we're dealing with a sort of plug-in data port, advanced but not yet wireless, "The white socket for her neural connector flashed briefly as the tap came off and she fluffed her hair into shape" (8). Data ports become commonplace in SF. The engineer Harper has one, for example, in the television series *Andromeda* (2000–2005). (Incidentally, the experimental mice are not yet wireless either.)

Pellegrino, Powell and Asimov's Three Laws of Alien Behavior

"Law No. 1 Their survival will be more important than our survival. If an alien species has to choose between them and us, they won't choose us. It is difficult to imagine a contrary case; species don't survive by being self-sacrificing. Law No. 2 Wimps don't become top dogs. No species makes it to the top by being passive. The species in charge of any given planet will be highly intelligent, alert, aggressive, and ruthless when necessary. Law No. 3 They will assume the first two laws apply to us" (Pellegrino 1).

In *Flying to Valhalla* (1993) and *The Killing Star* (Pellegrino and Zebrowski, 1995) Charles Pellegrino argues the proposition that spacefaring (or potentially spacefaring) civilizations will wipe each other out on principle. This moves considerably beyond the idea of misunderstandings in a first meeting between alien species (See **First Contact**.) Pellegrino's point is that once a sentient race has a technology sufficiently advanced to make the destruction of other spacefaring (or about-to-be spacefaring) species possible, prudence and an obsessive desire for self-preservation will inevitably lead to the use of that technology.

In the two novels I've mentioned, the method of destruction is a spaceship capable of near-relativistic speeds, but enough asteroids, like the one the alien invaders use against Earth in Niven and Pournelle's *Footfall*, could also get the job done without all the trouble of making large quantities of antimatter. (See **Doomsday Machine**.) Unfortunately, the science involved here is (according to James Gunn in his Introduction to the Easton Press edition of *The Killing Star*), "As real as tomorrow's headlines" (x).

There are, of course, many people with different views, Clifford Simak especially notable among them. In his Introduction to *Skirmish: The Great Short Fiction of Clifford D. Simak* (1977), he wrote, "My reluctance to use alien invasion is due to the feeling that we are not likely to be invaded and taken over. It would seem to me that by the time a race has achieved deep space capability it would have matured to a point where it would have no thought of dominating another intelligent species. Further than this, there should be no economic necessity of its doing so. By the time it was able to go into deep space, it must have arrived at an energy source which would not be based on planetary natural resources" (ix).

Forty years later, there is some scientific support for Simak's optimism. **Nanotechnology** may be expected to transform planetary economies of scarcity into spacefaring economies of plenty. In *The Killing Star*, Pellegrino and Zebrowski write concerning the largest object in the asteroid belt. "More than twenty kilometers below the spreading dust fields, the first and largest of Ceres's inner shells was nearly complete. It resembled a giant bicycle wheel embedded in the asteroid, and it provided each of the homesteaders with a ranch larger than some Earth counties" (37). If a dwarf planet can provide that sort of real estate, and if, as now seems likely, the universe is overflowing with planets of various types, there would seem to be no ghost of a reason for spacefaring races to compete—certainly not to the death.

Now, I realize that to some extent I have set Pellegrino and Simak arguing at cross purposes. An attempt to seize resources or colonize a planet is a very different goal from eliminating a civilization because it could possibly pose an existential threat. George Zebrowski says in the Afterword to *The Killing Star*, "*Nothing* outweighs even the distant threat of genocide" (339). Still, I hope that Simak is correct in assuming "by the time a race has achieved deep space capability it would have matured to a point" where it would not consider annihilating other sentient species on the off chance that they might make trouble. Another consideration here is that not all annihilating terminations will succeed, and a retaliation for such a deadly blow might be expected to be a far greater threat than letting the creatures live in peace in the first place.

The unstated assumption behind these three laws is that the true resolution of the Fermi Paradox is the propensity of spacefaring races to wipe each other out. There are many other explanations. One put forward by George Zebrowski is that "most Earth-like planets have not been around long enough to sprout civilizations" (332). Or in the considerably more poetic phrasing of astrophysicist Mario Livio, "If civilizations exist around other stars they are likely to be just emerging across our Galaxy right now: like an apple orchard suddenly maturing and ripening in the autumn sun." (See **Fermis**.)

I'm going to leave George Zebrowski with the last word. He may sound slightly paranoid, but as Harold Finch from *Person of Interest* would no doubt remind us, "Only the paranoid survive" ("Wolf and Cub"). And that's on this planet. Zebrowski writes, "We must think about what we are doing, in the light of what is possible and probable. And that is the ... grim reality that distinguishes 'hard' science fiction from every other kind of writing: not that something is merely possible and interesting, or metaphorical ... but that it is possible, probable, even likely; that it does not exist only in the imagination, but might confront us in our daily lives—as for example an asteroid strike that might destroy most of civilization" (340).

Positronic Brain

From the beginning, Isaac Asimov's robots were capable of thinking as well or even slightly better than their human counterparts. It was the positronic brain that made this believable. Asimov coined the term for "Reason," his second robot story, which appeared in *Astounding Science Fiction* in April 1941. Technically, "Reason" is the third story in the sequence, which is where it appears in *I, Robot*, Asimov's collection of his early robot stories, but publication dates put it second.

"Reason" is a Powell and Donovan story, two troubleshooters whose job it is to find and fix the bugs in new-model robots. Though Asimov used the term in the story's first paragraph, calling it "an inscrutable positronic brain" (227), it's not until Powell and Donovan begin assembling a robot that we get the details: "It was the most complicated mechanism ever created by man. Inside the thin platinum-plated 'skin' of the globe was a positronic brain, in whose delicately unstable structure were enforced calculated neuronic paths, which imbued each robot with what amounted to a pre-natal education" (238–239). It was, as the narrator says in *I, Robot*, "about the size of a human brain" (7).

Asimov explained his "invention" of the positronic brain many times over the years. In "My Robots," he wrote, "Since I began writing my robot stories in 1939, I did not mention computerization.... The electronic computer had not yet been invented and I did not foresee it. I did foresee, however, that the brain had to be electronic in some fashion." But he needed something more "futuristic." Fortunately, "the positron—a subatomic particle exactly like the electron but of opposite electric charge—had been discovered only four years before.... It sounded very science fictional indeed, so I gave my robots 'positronic brains' and imagined their thoughts to consist of flashing streams of positrons, coming into existence, then going out of existence almost immediately" (454).

Positrons had initially been called antielectrons, or to put it another way,

114 Railgun

they were the first particles of antimatter to be discovered, and robots whose minds were made from them would certainly have thoughts as fleeting as moonlight in dreams. Paul Dirac had worked out in 1928 in his famous equation "the existence of antiparticles and in some sense 'foretold' their experimental discovery" (Pickover 350). In 1932, Carl David Anderson "detected a positively charged electron," the antielectron or as Anderson decided to call it the positron. "It was plain that, just as Dirac had predicted, the antielectron did not last long. Within a billionth of a second or so, it would encounter an electron, and mutual annihilation would take place" (Asimov "Opposite!" 259–260).

Though Isaac Asimov's robots represent a prediction that has come true, and **Asimov's Three Laws of Robotics** may yet prove the blueprint (or at least the thought experiment) that leads to the construction of intelligent and safe **artificial intelligence**s, the creation of the positronic brain is unlikely to prove equally prescient even though it has been the inspiration for a variety of other SF robots, whose creators (from *Star Trek*'s Data to the Daleks in *Doctor Who* to Twiki and Crichton in *Buck Rogers in the 25th Century*) have blithely appropriated the term.

Nevertheless, it may well clarify the goal even if it was never meant to be a realistic way to reach it. In an article titled "DARPA Working on Sentient Robots with 'Positronic' Brains," Gene Turnbow says, "A group of researchers in Malibu, California and the University of California, Los Angeles have built a tiny machine that would allow robots to act independently. This sounds very like Isaac Asimov's ideation of a 'positronic' brain, the foundation of artificial intelligence in his popular series of 'robot' novels."

There is, of course, no indication that actual positrons are involved. What's happening is that Asimov's goal of a mechanism that emulates the human brain is nearer than it has ever been. Turnbow says, "What sets this new device apart from any others is that it has nano-scale interconnected wires that perform billions of connections like a human brain, and is capable of remembering information.... Each connection is a synthetic synapse." Perhaps someday soon another of Asimov's ideas will reach reality. It will not be in the way he imagined, but very possibly it will do what he dreamed it would.

Railgun (also sometimes Mass Driver)

The *OED* defines it as "a device for accelerating particles or launching projectiles by accelerating them electromagnetically along a pair of electrically conducting rails" and gives its first example from *Science* in 1960. There is clearly some overlap between railgun and the broader term mass driver, which the *OED* defines as "an electromagnetically driven launching system, proposed as a method of propelling objects into space or over long distances."

Their first example is from futurist Gerard K. O'Neill in 1975 Congressional hearings, but their definition of mass driver is too narrow. Charles Sheffield calls his space elevator in "Skystalk" (1979) a "mass driver system" (192). David Brin says, "The notion of gun-propelled launch goes back to Jules Verne. Such Mass Drivers have been envisioned in numerous Sci Fi tales, including *Earthlight*, by Arthur C. Clarke, Robert A. Heinlein's *The Moon Is a Harsh Mistress*, and *Heart of the Comet* by Benford & Brin" ("Space-Launch Mass Drivers").

Isaac Newton is credited with suggesting a space gun, though it might be closer to the truth to say he suggested a mass driver. In a thought experiment from his 1728 work *A Treatise of the System of the World*, he discusses what would happen if an object were "projected" from a tall mountain with greater and lesser force, "If the velocity was still more and more augmented, it would reach at last quite beyond the circumference of the Earth, and return to the mountain from which it was projected" (6). If the height were far greater, such projectiles could "go on revolving through the heavens in those trajectories just as the Planets do in their orbs" (7).

As usual, Jules Verne's suggestions seem more practical. In *From the Earth to the Moon*, Verne's indomitable Barbicane says, "I have looked at the question in all its bearings, I have resolutely attacked it, and by incontrovertible calculations I find that a projectile endowed with an initial velocity of 12,000 yards per second, and aimed at the moon, must necessarily reach it." As a result of this book, space guns are also called Verne guns. There is no question what Verne had in mind. Barbicane says, "The resisting power of cannon and the expansive force of gunpowder are practically unlimited."

Heinlein's mass driver is initially used to send grain from the moon to the Earth (and yes, I have that the right way round). When relations between the moon and Earth deteriorate, "Lunar Authority was invited to gather at one spot well away from other people, say in unirrigated part of Sahara, and receive one last barge of grain free — straight down at terminal velocity…. [W]e were prepared to do the same to anyone who threatened our peace, there being a number of loaded barges at catapult head." In the event, though, it's not grain they mean to send, "Greatest effort went into moving steel to new catapult and shaping it into jackets for new rock cylinders" (*The Moon* 295).

While mass drivers can have peaceful purposes (Barbicane's gun is a means of launching humans to the moon and NASA has worked on mass drivers as launch systems), they are often envisioned as weapons. John Munro's *A Trip to Venus* (1897) is a very early indication of something that sounds remarkably like a possible path to a railgun. In keeping with the usual close integration of science and science fiction, Munro wrote a number of nonfiction books about electricity. Here is his description of an electronic cannon, "We could even have an electric gun. Conceive a bobbin wound with insulated wire in lieu of thread, and having the usual hole through the axis of the frame. If a current of electricity be sent through the wire, the bobbin will become a hollow magnet or 'solenoid,' and a plug of soft iron placed at one end will be sucked into the hole. In this experiment we have the germ of a solenoid cannon."

There were soon real guns of extraordinary size such as the German "Paris" gun from World War I, "a grotesquely big cannon that could bombard Paris from about 75 miles away" and incidentally send its shells to the stratosphere so that "the Paris Gun operators needed to take into account the Coriolis effect (the rotation of the Earth)" (Patel).

The first true railgun was built in Australia as one of the experiments made possible by Sir Mark Oliphant's homopolar generator at Australian National University. It was one of a fascinating group of possibilities. Unfortunately, "for one reason or another, none of the projects was provided with sustained support" (Ophel and Jenkin 31).

Functioning railguns were finally produced by the United States Army and Navy in separate programs. Science fiction, which had, as this entry makes clear, been alternately predicting and benefiting from the ideas of mass drivers and railguns, continued its symbiotic relationship. In February of 2008, Brendan Borrell wrote, "At the Naval Surface Warfare Center ... a seven-pound bullet emerged from a truck-sized contraption at seven times the speed of sound and sent a visible shockwave through the air before crashing into a metal bunker filled with sand. With 10.6 megajoules of kinetic energy, this aluminum slug was propelled not by explosives but by an electric field, making this the most powerful electromagnetic railgun ever fired."

The previous record "of 9 megajoules had been set 15 years ago by a team at the University of Texas at Austin funded by the U.S. Army" (Borrell). In between those two dates (on March 25, 2005), *Stargate Atlantis* aired an episode titled "The Siege (Part 2)." Under attack by Wraith ships, Atlantis receives reinforcements from Earth. Colonel Everett announces, "We brought along a few railguns.... They will deliver an impact velocity of Mach 5 at 250 miles; a standard magazine will hold 10,000 rounds."

Railguns provide big advantages because they outperform explosive shells in both distance and destructive power without the disadvantages of transporting and storing gunpowder. A ship equipped with an effective railgun that was capable of firing continuously could conceivably destroy an entire fleet of conventionally armed vessels. The disadvantages of railguns are that they require tremendous power (more than most existing ships can generate) and that in their current stage of development, they destroy themselves very quickly as they fire.

In May of 2016, the U.S. Navy was testing a railgun with projectiles that traverse "the length of a 32-foot barrel, exiting the muzzle at 4,500 miles an hour, or more than a mile a second" (Barnes). To make a direct comparison with the *Stargate Atlantis* weapon, nearly Mach 6. Patrick Tucker reports it can "fire a shell at up to 5,600 miles per hour" or more than Mach 7. It "requires a power plant that generates 25 megawatts—enough electricity to power 18,750 homes." Power consumption at that level is part of the bad news, but "The Navy now believes it has a design that soon will be able to fire 10 times a minute through a barrel capable of lasting 1,000 rounds" (Barnes).

It's not clear, though, how things will work out. "The gun is supposed to be deployed on the new Zumwalt class destroyers since the Zumwalt is basically the only ship that generates enough juice, 78 megawatts, to power the gun. But the military is making just three of the ships" (Bennett "The Future").

"The Navy began working on the railgun a decade ago and has spent more than half a billion dollars. The Pentagon's Strategic Capabilities office is investing another $800 million" (Barnes). Research obviously will continue with the objectives of reducing power consumption and the size of the gun. Also important is increasing the gun's durability. One additional option that the Navy has found appealing is another way of using railgun projectiles. "In 2012, the Pentagon realized that they could fire the railgun's projectile out of the 5-inch powder guns on existing ships. No, it wouldn't hit speeds of Mach 7 (topping out closer to Mach 3), but that's twice as fast as a normal round fired from a 5-inch powder gun" (Tucker).

In 2015, P. W. Singer and August Cole's *Ghost Fleet: A Novel of the Next World War* explored the future of the railgun, Zumwalt class destroyers, and the relationship of the U.S. and China, among other things. Only one of the Zumwalt class destroyers is completed, and it's in the ghost or reserve fleet. Its technology is beset with problems, and at the climax of the novel, it is reduced to firing on the Chinese fleet while it is in danger of sinking, a problem the *Zumwalt*'s captain ignores in order to keep the gun in operation: "The steady explosions of the rail gun releasing rounds continued. One round every six seconds with a metronome's precision.... In the distance, small bright flashes and then black plumes began to appear, the only visual indicators of the steady rail-gun shots working their way through the enemy task force" (369–370).

Here is Singer and Cole's description of a round striking a ship. "The rail-gun round entered the *Admiral Zheng He*'s superstructure approximately thirty feet beneath where Admiral Wang stood. The strike transferred its kinetic energy with such force that the metal superstructure was literally peeled apart as the round plowed through. The ensuing explosion amidships sent a ball of flame hundreds of feet into the air as the ship's hull cracked in two" (369). (The inconsistency in spelling the term is in the original.)

It seems unlikely in the context of the novel and the real world that such a weapon, which may soon be able to hit incoming nuclear missiles as well as ships, will not become part of U.S. and other arsenals—unless an international agreement banning its use gets in the way.

Meanwhile, the real *Zumwalt* "was commissioned in Oct. [2016] in a ceremony at Baltimore" (Cavas). Like its fictional counterpart, the real ship has had its problems. It "has been dealing with a series of relatively minor incidents, including a seawater leak in a shaft lube oil system in September" and in November "an engineering casualty ... while passing through the Panama Canal "which meant it had to be towed.... The installation of its combat system" in San Diego is "a job expected to continue through most of 2017" (Cavas).

Though it does not yet sport a railgun, "the Zumwalt's powerful new gun system can unload 600 rocket-powered projectiles on targets more than 70 miles away" (Garucho). It has stealth capabilities and could be described as the most technologically advanced ship in the U.S. Navy. Vice Admiral Tom Rowden says, "The Zumwalt's unique and significant capability to generate power could be used in ways perhaps not even envisioned yet, such as in the testing and use of laser and directed-energy weapons systems" (Garucho).

One small, entirely coincidental note: the most technologically advanced ship in the U.S. Navy is commanded by a captain named James Kirk.

Realtime

If travel to the past is possible, there are three relatively plausible outcomes. The first (and, most people think, the most likely) possibility is that the past is fixed and everything that can happen has already happened. (See **Time Machine**.) By that argument, time travelers to the past will do what they have already done, and the past will not be changed.

The second possibility is that the past can be changed, and the future from which the time traveler came will be altered or even destroyed by those changes. (See **Timequake**.)

The third possibility is that time travelers can make changes, but those changes will not affect the timeline from which they came. Instead, the changes will create an alternate (alternative) or parallel timeline in which the time travelers may or may not be trapped. If this third outcome turns out to be the correct one, there will then be Realtime, the original timeline from which the traveler or travelers came as opposed to the other timeline or timelines created by the changes the time travelers make. These new timelines will, in all probability, be viewed by time travelers (especially those who can move among them) as less real than the original. This will be reinforced by the fact that history remains unchanged in the original timeline but is infinitely variable in the others.

It was first used by Bruce Sterling and Lewis Shiner in their 1985 story "Mozart in Mirrorshades." "You got to understand how the portal works. Right now it's as big around as you are tall, just big enough for a phone cable and a pipeline full of oil ... heading for Realtime." As the Thomas Jefferson character in that story describes the situation, "This world—my world—does not lead to your future.... Leaving you free to rape and pillage here at will! While your own world is untouched and secure!" (See **Time Bunny**.)

There is an identical interpretation of time travel in Steven Spielberg's *Terra Nova*, a 2011 television series, where a future society exploits a period eighty-five million years in the past because it is in an alternate timeline. In the first episode, it is called "a fracture in the fabric of time and space" and "a new time stream" ("Genesis Part 1"). There have been many other stories that used the idea of an alternate timeline split off from an original, without the added factor of exploitation, including the films that reset the *Star Trek* Franchise, *Star Trek* (2009), *Star Trek Into Darkness* (2013), and *Star Trek Beyond* (2016).

The idea of a central world with many parallel worlds or alternate timelines is, of course, instantly recognizable as Roger Zelazny's Amber series. *Nine Princes in Amber*, the first novel in the sequence, was published in 1970. Zelazny says, "If there is an infinity of parallel worlds in which anything can exist, and if one then allowed for a race of intelligent beings with the ability to traverse any of these worlds under their own power, then it follows that one particular world must be the keystone or archetypal world" (Krulik Zelazny). Beyond the basic concept, however, (some elements of which go back as far as Plato) there is not much common ground between "Mozart in Mirrorshades" and Zelazny's fantasy. (See *Archetype*.)

Robot

Machines that can do almost anything either as individuals or in groups. They may be reprogrammable, capable of learning or both, and their shapes and sizes vary to fit the tasks they are designed to perform. In SF, robots are usually mechanical copies of humans. The copy may only vaguely resemble the original, as with Robby in *Forbidden Planet*, or it may be so precise as to challenge detection, as with Daneel in the Asimov books. As this indicates, the dividing line between robot and android can be unclear.

Jeff Prucher's definition of robot shows the problem, "an intelligent or self-aware artificial being, especially one made of metal" (164). It might be possible to distinguish between robot and android by suggesting that at least some part of an android must be biological so that Data and the Terminator are clearly androids, but even that is problematic. In the *Logan's Run* television series, Rem, who seems completely mechanical, resents being called a robot, because the term is too primitive to apply to him, "Call me an android ... what you will, but please never ever call me a robot.... I'm ... the ultimate computer in human form" ("Logan's Run").

The original distinction of computer (or brain), robot, and android comes from Edmond Hamilton's Captain Future series: "Simon, the Brain that long ago had lived in a man's skull but now lived in a cubical case" (15). "Grag the robot, the metal giant of the Future-men!" (14). "And the android, most manlike of the three, human in all but origin" (15).

The idea of a robot (and an ASI— see **Artificial Intelligence**) existed more than 2500 years before the word came to express it. In Book 18 of Homer's *Iliad*, Hephaestus (Vulcan) the "lame" god of the forge, is helped and supported by robot maidens, "He donned his shirt, grasped his strong staff, and limped towards the door. There were golden handmaids also who worked for him, and were like real young women, with sense and reason, voice also and strength, and all the learning of the immortals; these busied themselves as the king bade them, while he drew near to Thetis" (293).

Not all of Hephaestus' robots are humaniform. Even here, in their first appearance in literature, they come in different shapes and perform separate functions; some are capable of feats beyond the grasp of humans and some are little more than animate furniture, "He was making twenty tripods that were to stand by the wall of his house, and he set wheels of gold under them all that they might go of their own selves to the assemblies of the gods, and come back again—marvels indeed to see" (292).

"Automaton" was the most frequently used word before "robot." For instance, in 1893, in Ambrose Bierce's short story "Moxon's Master," an "automaton" strangles its maker when he beats it at chess (long an area of contention between machines and humans—see **Deep Blue**). Here is how the *OED* defines the term that quickly became universal: "Chiefly *Science Fiction*. An intelligent artificial being typically made of metal and resembling in some way a human or other animal. Originally with reference to the mass-produced workers in Karel Čapek's play *R.U.R.: Rossum's Universal Robots* (1920) which are assembled from artificially synthesized organic material."

Though Karel Čapek is usually listed as the inventor of the term, he gave the credit to his older brother, Josef, cubist painter, illustrator, and writer. As Karel tells the story, he had an inspiration for a play but was dissatisfied with the word for the creatures. He rushed to his brother for advice, telling him he had an idea for a play: "'What kind,' the painter mumbled (he really did mumble, because at the moment he was holding a brush in his mouth)." Karel detailed the plot briefly. "'Then write it,' the painter remarked, without taking the brush from his mouth or halting work on the canvas. The indifference was quite insulting. 'But,' ... 'I don't know what to call these artificial workers. I could call them *Labori*, but that strikes me as a bit bookish.' 'Then call them Robots,' the painter muttered, brush in mouth, and went on painting. And that's how it was. Thus was the word Robot born" ("Who Did Actually Invent").

Some people have suggested that the word did not emerge around a paint brush but was first published in Josef Čapek's short story "Opilec" ("The Drunkard") in *Lelio*, published in 1917. In fact, the word used in the story is "*automat*" ("Who Did Actually Invent").

On October 9, 1922, Karel Čapek's play *R.U.R.* opened in New York, introducing the word "robot" (from the Czech *robota*, meaning work or compulsory service) to America. They eventually revolt, destroying their creators (and enslavers). (See **Frankenstein Complex**.) A robot Adam and Eve appear from the originally sexless creatures so we may assume they survive. The production was comparatively successful, running for 184 total performances and closing the following February. Along the way, it provided roles as robots for Pat O'Brien and Spencer Tracy in their Broadway debuts.

Ironically, the word for one of the most important ideas in science fiction and not long after that, science fact, was popularized by someone who, according to Majer and Porter, in his "later writings criticised the tyranny of reason and the heartlessness of modern technical civilisation, by focusing on some new scientific or technical discovery which he imagined as dehumanising and interfering with the world and driving it to its destruction" (x).

As Adam Roberts points out, "The company title 'Rossum' is a play on the Czech word *rozum*, which means 'reason' or 'intellect' (168). In fact, Majer and Porter translate the play's title as *R.U.R., or Reason's Universal Robots*. Roberts says, "The set-up is only too evidently that of a hypostatized 'mind/body,' or 'masters/workers' binary" (108). In that sense, it is not unlike H. G. Wells' Eloi and Morlocks in *The Time Machine*, and that suggests a contemporary application rather than a prediction of the future.

If *R.U.R.* was meant to be a self-defeating prophecy, it was an abysmal failure. The idea behind "robot" (at least the idea as interpreted by Isaac Asimov and similar techno-optimists) and the word itself began—as so often happens—to change the real world. In 1950, Isaac Asimov's first collection of stories about good robots, *I, Robot*, was published. Asimov said, "As a machine, a robot will surely be designed for safety as far as possible. If robots are so advanced that they can mimic the thought processes of human beings, then surely the nature of those thought process will be designed by human engineers and built-in safeguards will be added" (*The Rest of the Robots* xiii).

"Joseph Engelberger, who built the first industrial robot, called Unimate, in 1958, attributes his long-standing fascination with robots to his reading of *I, Robot* when he was a teenager" (Asimov, Warrick, Greenberg 69). That is one of history's fastest turnarounds from inspiration to implementation.

"In 1959, General Motors was the first to install a test model Unimate in its Turnstead die-casting plant. But orders for more Unimates did not come until 1961 (Asimov and Frenkel 33). Few people were more committed to the use of the word "robot" than Joseph Engelberger. "Over and over, the advice was 'Don't call it a robot. Call it a programmable manipulator. Call it a production terminal or a universal transfer device.' The word is *robot* and it should be *robot*. I was building a robot, damn it, and I wasn't going to have any fun in Asimov terms, unless it was *robot*. So I stuck to my guns" (Asimov and Frenkel 25).

Futurists, scientists and science fiction writers all began to speculate about what computers and robots might be able to do. In 1950, in his essay "Can a Machine Think?" Alan Turing proposed the Turing test to determine whether a machine had reached a human level of intelligence.

In 1959, Richard Feynman, in the same Caltech address to the American Physical Society where he suggested **nanotechnology**, "talked about computers: given millions of times more power, they would not just calculate faster but would reveal qualitatively different abilities, such as the ability to make judgments. 'There is nothing I can see in the physical laws that says the computer elements cannot be made enormously smaller than they are now,' he said" (Gleick 355).

In June of 1960 Clifford D. Simak's "How-2" was reprinted in *Bodyguard*, an anthology of "short science fiction novels from *Galaxy*" (an important SF magazine published between 1950 and 1980). "How-2" concerns an experimental, accidentally released (and assembled) "mother robot" named Albert, who makes other robots, which s(he) considers to be her children. As the result of a major legal case (with Albert's specially designed, telepathic legal robots providing the research), Albert and other robots are declared "people," and humans are left with a sense of almost complete uselessness, a feeling that has been growing on them since, as a result of machine efficiency, the work week was reduced to fifteen hours (three five-hour days). Now, even the How-2 hobbies humans had taken up to make themselves feel useful will be usurped by robots, who plan to make already assembled How-2 kits, saving humans yet more time. This is obviously not a positive picture of the future of robots, but it raised an extraordinarily important issue.

On February 2, 1961, William Welch's comedy-drama *How to Make a Man* premiered on Broadway (at the Brooks Atkinson Theatre). It was based on Simak's "How-2," and with its court case involving the rights of robots, it was a play ahead of its time. Welch had written scripts for most of Irwin Allen's SF television series, and Allen considered doing *How to Make a Man* as a television series also, but Welch died before the project could be realized.

Later explorations of the issue include Isaac Asimov's book *The Bicentennial Man* published in 1976. (There is in addition a 1993 novel written with Robert Silverberg, *The Positronic Man*, and a film in 1999.) In the title story, a robot gradually has his metal parts replaced with human ones until he is completely human, at least in materials. Along the way (but before he has started changing his form), he is granted human rights. The judge decides, "There is no right to deny freedom to an object with a mind advanced enough to grasp the concept and desire the state" (144).

Data goes on trial for his life (or at least for the preservation of his coherent identity and personality) in the *Star Trek: The Next Generation* episode "The Measure of a Man," first aired on February 11, 1988. It too is a precedent-setting case about the legal rights of artificial humans. One question that all of these fictions raise is how soon will there be a real case along the same lines?

The definition of robot is continuing to evolve, just as robots themselves are doing. On one side, Harry Waldman's *Dictionary of Robotics* had managed a fairly stable, nearly universal definition of one kind of robot by 1985: "A mechanical device that can be programmed to perform tasks of manipulation and locomotion under automatic control. A reprogrammable manipulator that performs functions normally done by a person, working in essentially the same manner as does a human but at a faster speed" (216). This is, of course, the definition of an industrial robot. Such things are steadily becoming cheaper, more flexible and more productive.

A very different kind of robot (entirely outside that definition but within mine) is also at work. "Warehouses run by Gap, as well as Zappos and Staples now use autonomous robots to pluck products from their shelves and send them to you. All the robots are told is where products are located and where they need to go. From there, the robots, which look like massive orange Roombas, figure out the rest. They locate the stack of shelves with the needed product on it, slide beneath the stack to pick it up and then find their own routes from the stacks of stuff to human operators" (Madrigal). Robots do a wide range of other jobs, and their accomplishments increase every day. Personal robots are not too far away, and "self-driving" cars and other vehicles may soon be commonplace.

The underlying questions are still here. In a 1966 episode of *Lost in Space* called "The Wreck of the Robot," the robot explains his success in thwarting the villain. "He said that their invention could not be destroyed by neither man nor machine [sic]. It is obvious that I am not a man, but I am also not quite a machine. I suppose I must be something in between."

Indeed, robots seem to be in between and on the verge of many things. Will they become like Asimov's creations, as smart as humans but no smarter? Will they rapidly surpass unmodified humans, reaching a level of intelligence beyond what we can presently imagine? Or will they remain nothing more than the mechanical means to our dreams of reliable servants who will do all our work, make us rich by their labor and even fight our wars for us? At this moment, there is no obvious answer.

Robotics

A word created when Isaac Asimov and John W. Campbell came up with **Asimov's Three Laws of Robotics**. It was in the March 1942 *Astounding* in a story titled "Runaround," as Asimov just happens to remember, "on page 100, in the first column, about one-third of the way down" (*Robot Visions*, 7). In *The Rest of the Robots*, Asimov calls robotics "a non-existent science" (43), but that was in 1964. Things (and robots) changed quickly.

The Random House Dictionary of the English Language: The Unabridged Edition, first published in 1966 and republished until 1981, does not contain the word. But by 1985, there was Harry Waldman's *Dictionary of Robotics* and also, as Asimov points out in *Robot Visions*, "a huge tome, *Handbook of Industrial Robotics*, edited by Shimon Y. Nof" (456–457). Whether it was as a direct result or not, the 1993 *Random House Unabridged Dictionary* included the word, defining it as "The use of computer-controlled robots to perform manual tasks, especially on an assembly line" (1664).

I would add a second definition: the science of designing, making or dealing with robots in any of their many forms. Undoubtedly, the meanings of the word will continue to grow and become more complex as robots themselves do.

SETI

An acronym made from the Search for ExtraTerrestrial Intelligence. The idea has a long history; the notion of searching for radio signals or other electronic sounds goes back at least as far as Nikola Tesla. Frank Drake is generally credited with kicking off modern experiments with Project Ozma in 1960. Even though the *OED* doesn't find the first use of the term SETI until 1976, many ideas and a wide range of books preceded it.

One of the most important books (and one of the earliest) was *Intelligent Life in the Universe*, written by Soviet astrophysicist Iosif Shklovsky in 1962 and then expanded and co-written by Carl Sagan in 1966. (Sagan's 1985 SF novel, *Contact*, is an extension of his fascination with the subject.) While my first example is by astrophysicists (one of whom would eventually become a science fiction writer), my second example was written by an anthropologist (and poet) who was about to emerge as a futurist. In 1970, Loren Eiseley wrote *The Invisible Pyramid: A Humanist Account of the Space Age*. The book looked deep into the human past and also speculated on the future of humans and other spacefaring species.

For my third and final example, I mention James Gunn's 1972 SF novel *The Listeners*. As usual, science, futurism, and SF work together to make something new. Gunn credits Walter Sullivan's *We Are Not Alone* as one of his inspirations, and toward the end of his new (2000) Preface to his own book, he writes, "One of the SETI project directors told me recently that *The Listeners* had done more to turn people on to the search than any other book."

Discussing the search for extraterrestrial intelligences in his *Dictionary of Scientific Literacy* (1992), Richard P. Brennan said, "One approach has been the

NASA-sponsored project that listens for artificially generated electromagnetic signals coming from space on the assumption that any technologically advanced civilization would be producing radio/TV/radar signals just as Earth does" (269). Allowing for advances in our ability to listen to and sort through possible messages since 1992 and a far more efficient system for finding exoplanets, that is a workable definition of passive SETI, a search for aliens that might allow us to find them without signaling our own existence.

Active SETI (also called METI, an acronym for Messaging to Extraterrestrial Intelligence) argues as Alexander Zaitsev, "Chief Scientist at the Russian Academy of Sciences' Institute of Radio Engineering and Electronics," does, "that we as a species have a moral obligation to announce our presence to our sentient neighbors in the Milky Way— to let them know *they* are not alone. If everyone in the galaxy only listens, he reasons, the search for extraterrestrial intelligence (SETI) is doomed to failure" (Grinspoon).

Zaitsev set off a firestorm, or at least an editorial in *Nature*, and, more importantly, some resignations from the "Permanent Study Group ... of the SETI subcommittee of the International Academy of Astronautics, a widely accepted forum for devising international SETI agreements." He did that by sending "several powerful messages to nearby, sun-like stars" without consulting anyone (Grinspoon).

There are several areas of disagreement here. First, many members of this community of experts believe that there should be carefully monitored group decisions (including input from non-specialists) before any such steps are taken. As David Brin puts it, "We have resigned from various SETI-related international committees in protest over those who have rushed to beam 'Yoo-hoo' shouts out into the cosmos without ever exposing their concepts to collegial or peer review" (cited in Helen Mann). Second, and this is where the real dispute comes in, there is the question of whether we should be talking to space aliens at all. To quote David Brin again, "It is this rush to say, 'Of course we know exactly what we're doing' just like in a Michael Crichton movie. That we're assuming that the universe can only be altruistic, the universe can only be benign" (cited in Helen Mann).

And in an issue this complex, it's not only the nature of the universe that can be confusing. Stephen Hawking, for instance, has been (mostly) on the other side of the question, declaring aliens to be dangerous. At the same time, he has partnered with Russian billionaire Yuri Milner in a project called "Breakthrough Listen" to "spend at least $100 million in the next decade to search for signals from alien civilizations." It sounds as though it will be the most extraordinary example of passive SETI ever attempted.

"It will allow astronomers to see the kinds of radar used for air traffic control from any of the closest 1,000 stars, and to detect a laser with the power output of a common 100-watt light bulb from the distance of the nearest stars, some four light-years away" (Overbye). And yet Hawking sometimes sounds as though he's a commentator discussing events from inside a real-world *Independence Day*, where the only responsible way of saying Welcome to Earth is with a fist in the face.

In *Stephen Hawking's Universe* (2010) he said of aliens, "They might exist in massive ships, having used up all the resources from their home planet.... Such advanced aliens would perhaps become nomads, looking to conquer and colonize whatever planets they can reach" (cited in Baral). More recently, in a half-hour program on Curiosity Stream (2016), he repeated his warnings, "One day we might receive a signal from a planet like this, but we should be wary of answering back. Meeting an advanced civilization could be like Native Americans encountering Columbus. That didn't turn out so well" (cited in Baral).

Hawking to the contrary notwithstanding, the real trouble with space aliens is unlikely to be that they will exceed the worst excesses of their fictional counterparts in the television miniseries *V*, or even in Anne McCaffrey's *Freedom's Landing* or Christopher Anvil's *Pandora's Legions*, to take a couple of less well known examples of aliens behaving badly. The true existential threat may well be found in the kind of paranoia that Pellegrino and Powell explore in *Flying to Valhalla*. They demonstrate what might happen if an alien race were to believe that any spacefaring species is potentially dangerous to every other species and, therefore, should be destroyed to eliminate the threat, however small.

One explanation for Fermi's Paradox (the strange absence of other intelligent creatures in the universe) is that as soon as a new spacefaring species is identified, it is exterminated. (See **Fermis** and **Doomsday Machine**.) Brin points out, "An alien probe, resident in our solar system, might be commanded to steer asteroids toward Earth. Or else to meddle in our computer networks or in our political affairs. The list of unlikely but physically quite possible scenarios is very long, many of them already explored in cogent, higher-end science fiction tales, a great library of thought experiments that SETI-ists seem willfully bent on ignoring" (Brin "The Search").

There are, of course, a host of other possible dangers, many of them beyond even the capacity of SF writers to imagine because they involve creatures who, even if they are at a technological level approximately similar to our own, will very likely be beyond our understanding as a result of their wildly different origins and development. And if their evolution and technology are significantly more advanced, it may be that we will not be smart enough to communicate with them at all. Indeed, perhaps even now we are being bombarded by messages that seem to their senders to be simple and straightforward, sent in a way that any sapient species could grasp, and yet we have less chance of detecting them than a raven has of deciphering Wi-Fi.

Still, there is a great longing in humans for other minds, and it is unlikely we will stop looking. Loren Eiseley said in 1959 in an essay titled "Little Men and Flying Saucers," "In a universe whose size is beyond human imagining, where our world floats like a dust mote in the void of night, men have grown inconceivably lonely.... As the only thinking mammals on the planet—perhaps the only thinking animals in the entire sidereal universe—the burden of consciousness has grown heavy upon us. We watch the stars, but the signs are uncertain" (161). Or as James Gunn wrote in *The Listeners*, "Back behind everything, lurking like a silent shadow behind the closed door, is the question we can never answer except positively: Is anybody there?"

Solar Sail

Using a sail in space driven by photons (or sunlight and therefore also called a light sail or space sail) in somewhat the same way that wind sails are used at sea. Plastic, aluminized mylar, another polymer, or a more exotic material would make up the sail, which would have to be extremely thin to minimize weight and extremely large—around one square mile according to some estimates—to maximize power.

The Handy Space Answer Book (1998) suggests that such a sail could produce a speed of 13 miles per hour on the first day. "One day later it would be moving at 200 miles (322 kilometers) per hour; and after eighteen days its speed would reach 1 mile (1.6 kilometers) per second while still picking up speed. At that rate of acceleration, it would take about 400 days to reach Mars" (Engelbert and Dupuis 413).

Another possibility, known as beam sailing, would be to use lasers for power. Imagine a solar power facility on the Moon with laser beams focused on the vast sails of ships moving through the solar system and perhaps beyond. Or imagine, as Niven and Pournelle put it in *The Mote in God's Eye*, "a battery of laser cannon" (66).

Astronautical engineer Robert Zubrin (in *Entering Space: Creating a Spacefaring Civilization*) has a shorter definition of solar sail than mine but workable just the same, "A device for propelling a spacecraft by utilizing the pushing force of sunlight" (289). The *OED* defines it as "a surface designed to utilize the pressure of solar radiation to provide the propulsive force for a spacecraft to which it is attached," but that is, perhaps, unclear. There are two types of radiation from the sun, the solar wind and sunlight, and only the second of the two is usable for driving a solar sail.

Isaac Asimov (with support from Arthur C. Clarke) says in *Project Solar Sail*, "The second 'wind' in space is the sun's torrential output of light, which also possesses momentum, and which exerts a miniscule pressure on anything it strikes.... As it turns out, sunlight itself is more than a thousand times stronger than the solar wind" ("Introduction" 4).

The distinction between the two types of radiation is clarified in Niven and Pournelle's *The Mote in God's Eye* (1974). The character Kevin Renner, who is plotting a course to deal with an alien craft powered by a solar sail, says, "The stellar wind can also propel a light sail.... The important difference is that the stellar wind is atomic nuclei. They stick where they hit the sail. The momentum is transferred directly—and it's all radial to the sun." And that leads his captain, who has no experience with solar sails, to a realization, "You can't tack against it.... You can tack against the light by tilting the sail, but the stellar wind always thrusts you straight away from the sun" (59–60).

Alas, we haven't yet been able to test these space sailing fictions against the facts of astronauts operating solar sails. However, Louis Friedman (in his 1988 nonfiction book *Starsailing*) has the methods neatly laid out, "The solar wind plays no part in solar sailing....

The solar sail operates on sunlight pressure—the pressure produced by light when it bounces off a mirror. This force is 1,000 to 10,000 times greater than that of the solar wind" (17).

He's also very clear about how tacking works, starting with the idea that any vessel moving through the solar system has orbital velocity, and "change of direction comes about by adding to, or subtracting from, orbital velocity" (20). If sunlight pushes straight against the sail, then velocity increases. We can "tack (change direction and, in particular, go inward toward the sun) ... by changing the angle of the solar sail with respect to the sun's direction so that the sail catches the particles of light on a slant." That creates "a force on the vehicle perpendicular to the sail. The sunlight pressure bounces off a solar sail at an angle, just as the wind bounces off a terrestrial sail, modifying the vehicle's velocity so as to produce changes in the orbital distance and direction" (20).

The *OED* gives 1960 as the first use of solar sail and the periodical *Aeroplane* as the publication. The second use was supposedly 1978 in *Listener*. However, the examples are incomplete. For some reason, the idea of the space sail or light sail or solar sail is one of the lacunae in the *OED*'s massive word hoard. It does not have space sail or light sail (or sailing in either case), and it ignores some of the contributions of SF writers to solar sail. It does have "solar sailing" but again gives 1960 and *Aeroplane*, followed by Carl Sagan in 1973. In the paragraphs that follow, I have a first use of space sailing in 1929 and Arthur C. Clarke using solar sailing in 1964. I will cheerfully admit that it's easier to write a book about words and the ideas behind them (or perhaps more accurately, ideas and the shifting words that swirl around them) than it is to deal with the massive problems of a historical dictionary.

The idea of a solar sail or space sail itself is, of course, far older than 1960. "In a 1608 letter to Galileo Galilei, Kepler wrote that humans might one day use the technology to set a course for the stars: 'Provide ships or sails adapted to the heavenly breezes, and there will be some who will brave even that void'" (Davis). As a result of his study of what would eventually be called Halley's Comet, Johannes Kepler concluded "that the tail of a comet appeared to be created by some sort of cosmic breeze and postulated that this breeze could be used to move ships in space" (Johnson). It was, for the time, a remarkable insight. "While Kepler was wrong about the nature of these cosmic winds, he was correct in his observation that something coming from the Sun, which we now know is sunlight itself, can be used to move a spacecraft" (Johnson).

Tennyson's words in "Locksley Hall" (written in 1835) have sometimes been taken to mean airplanes and sometimes solar sails, "Saw the heavens fill with commerce, argosies of magic sails/ Pilots of the purple twilight, dropping down with costly bales" (173).

There is little doubt what Jules Verne had in mind in *From the Earth to the Moon* (1865). "This will be exceeded some day by higher speeds still, of which light or electricity will be probably the mechanical agents." Konstantin Tsiolkovsky's "The Spread of Man in Space" (unpublished notes from the 1920s) has "possible methods of developing cosmic velocities," including "the pressure of light rays" (321).

In 1929, John Desmond Bernal, one of history's more extraordinary scientists and futurists, wrote *The World, the Flesh and the Devil*. In what was his own version of the **Cosmography of the Future**, he predicted that human beings would live in hollowed-out asteroids, creating an entirely new kind of society in the process. He also suggested, "A form of space sailing might be developed which used the repulsive effect of the sun's rays instead of wind. A space vessel spreading its large, metallic wings, acres in extent, to the full, might be blown to the limit of Neptune's orbit. Then, to increase its speed, it would tack, close-hauled, down the gravitational field, spreading full sail again as it rushed past the sun."

The idea of the solar sail, like so much of science and science fiction, is about adventure and imagination and even metaphor. We dream before we build, and the dream of space flight has always been as glittering as the night skies themselves. Carl Sagan said in his *Cosmos* series, "Exploration is in our nature. We began as wanderers, and we are wanderers still. We have lingered long enough on the shores of the cosmic ocean. We are ready at last to set sail for the stars" (193).

The images that equate the alien vastness of the ocean with the even more alien and far greater vastness of outer space have always seemed natural, and the extension of exploration from this world to an infinity of others has seemed natural as well.

Into this mix of metaphors and dreams, science and technology, solar sails fit perfectly since they are made up of equal parts of all of them. In his 1960 story for *Galaxy Magazine*, "The Lady Who Sailed the Soul," Cordwainer Smith (Paul Linebarger) said, "Out of it all, two things stood forth—their love and the image of the great sails, tissue metal wings with which the bodies of people finally fluttered out among the stars" (43).

In 1964, in Arthur C. Clarke's story "The Sunjammer" (later changed to "The Wind from the Sun" to avoid confusion with Poul Anderson's story), the main character thinks of "all his attempts to explain solar sailing to those lecture audiences" (16). But the story is also about the beauty of the endeavor and, as these stories almost always are, about the sea. The first words are, "The enormous disk of sail strained at its rigging, already filled with the wind that blew between the worlds" (15). And a little later, "The immense sail was taut, its mirror surface sparkling and glittering gloriously in the sun ... it seemed to fill the sky. As it well might—for out there were fifty million square feet of sail, linked to his capsule by almost a hundred miles of rigging. All the canvas of all the tea-clippers that had once raced like clouds across the China seas, sewn into one gigantic sheet, could not match the single sail that *Diana* had spread beneath the sun" (15).

It's hard to get away from the excitement of solar sails. Here's Dauna Coulter's definition of the term from "A Brief History of Solar Sails," "A gossamer material that, when unfurled in the vacuum of space, feels the pressure of sunlight and propelled by said pressure may carry a ship among the stars."

Paul Gilster writes in "Science, Fiction and the Sail," "The early story of the solar sail is inseparable from science fiction. *Astounding Science Fiction*'s John

Campbell published the first serious look at solar sails for propulsion all the way back in May of 1951. The article's title, 'Clipper Ships of Space,' would be echoed by a highly influential paper by Gregory Matloff and Eugene Mallove called 'Solar Sail Starships: The Clipper Ships of the Galaxy,' which ran in the *Journal of the British Interplanetary Society* in 1981." In his introductory comment to the essay in *Astounding*, Campbell called it "A fascinating suggestion for sailing ships of space, a seemingly wild notion, but worked out mathematically, it makes sense!" (in Saunders 136).

The list of science fiction writers, scientists and futurists who have written about and supported the idea of solar sails is extensive and distinguished. It included Pierre Boulle, whose 1963 novel *Planet of the Apes* begins and ends with two chimpanzees on a solar sailing holiday, "Their ship was a sort of sphere with an envelope—the sail—which was miraculously fine and light and moved through space propelled by the pressure of light-radiation" (9). One of the more remarkable results of the support from that community was *Project Solar Sail*, which was put together in 1990 to help the World Space Foundation in research for solar sails. The book was edited by Arthur C. Clarke, David Brin was managing editor, Isaac Asimov wrote the Introduction, and stories and articles were provided by, among others, Clarke, Brin, Poul Anderson, Ray Bradbury, K. Eric Drexler, Robert L. Forward, Larry Niven, and Charles Sheffield. It was one of several examples of privately funded groups such as the Planetary Society working to make the dream of the solar sail real.

In spite of plans by NASA and others for more than two decades, no solar sail was actually deployed in space until August of 2004. Japan's Institute of Space and Astronautical Science "deployed two 30-foot-wide solar sails, each less than 1/3000 of an inch thick.... At an altitude of 75 miles, one sail unfolded like a clover leaf. Then, at 100 miles, the other unfolded like a Japanese fan.... The Planetary Society (and Cosmos Studios) in the United States and private Russian space organizations developed Cosmos 1, which was meant to be the first fully functional solar sail. Unfortunately, the launch, on Tuesday, June 21, 2005, was unsuccessful, due to the malfunction of the Volna booster rocket" (McNicol 63).

It took until June of 2015, but the Planetary Society finally got its test sails into space. Loren Grush wrote, "Light-Sail has finally seen the light. After a tumultuous time in lower Earth orbit, the Planetary Society's solar sail spacecraft has at last deployed its namesake: four large mylar 'sails' that theoretically use energy released from the Sun to propel the satellite forward." The goal for this launch was no more than that—for the sails to open in space. The next step is scheduled for 2017.

Two solar sail spacecraft, however, had already done better. On January 20, 2011, "NASA's NanoSail-D spacecraft ... unfurled a gleaming sheet of space-age fabric 650 km above Earth, becoming the first-ever solar sail to circle our planet" ("Solar Sail Stunner"). NanoSail-D orbited the Earth for 240 days, which was longer than expected, burning up "during reentry" on September 17, 2011. Part of the purpose of the mission "was to demonstrate and test the deorbiting capabilities of a large low mass high surface area sail" ("NASA's Nanosail-D"). Or in non–NASA English, learn more about using solar sails to bring down old satellites and other space junk.

The prize for solar sailing through January 2017 (and mention in the *Guinness Book of World Records*) goes to Japan's IKAROS spacecraft which "deployed a solar sail in interplanetary space and used it to fly by Venus in 2010" ("Solar Sail Stunner"). IKAROS (more often spelled Icarus in English, the name of the son of Daedalus who flew too near the sun) is also an acronym for Interplanetary Kite-craft Accelerated by Radiation Of the Sun. The spacecraft flew by Venus, went on to the far side of the sun, and is now "sleeping" to store energy and then "waking up" to communicate with Earth.

Of particular interest is the structure of the IKAROS solar sail. "The sail, shaped somewhat like a kite, also has solar cells embedded to generate electricity. It was not expected to make much power during this flight, but more to serve as a test bed for future ion propulsion-engines that not only use solar cells for electricity, but also are moved along by sails" (Howell). Another successful experiment by the spacecraft was in attitude control, "through using liquid crystal panels at the fringes of the sail. Electricity flowing through liquid crystal panels gives a reflection for the spacecraft to move forward, while turning the flow off makes the sunlight pressure more diffuse. By using the different forces of sunlight on the edges, Ikaros was able to change orientation" (Howell).

It may be that solar sails are ready to join the **ion drive** (literally if the Japanese plans for a voyage to Jupiter work out) on the list of propulsion systems long written about in SF, anxiously awaited to help explore the real solar system, and now finally, successfully here.

A last note: "One of the most promising propulsion methods that might get us to the stars is photonic propulsion, or using a big laser to propel a small reflective spacecraft that has already been launched into orbit. This is how Stephen Hawking and Russian billionaire Yuri Milner hope to reach Alpha Centauri in our lifetimes as part of the Breakthrough Starshot initiative" (Bennett "Miniature 'Chipsats'").

The idea is that a laser might push a spacecraft, especially a very small one, "to very high velocities, as much as 20 percent of the speed of light, or about 100 million mph." The result could be a group of very small spacecraft "about the size of a sticky note, consisting of little more than a few basic electronics" reaching Alpha Centauri, the nearest star to the sun, in about twenty years. It's still an idea more than a plan, but as solar sails are even now demonstrating, ideas have a way of struggling through to reality.

Solo Parenting (also solo reproduction)

A human reproduction technique through which a single individual becomes the only biological parent to a child. The term is usually used in conjunction with **multiplex parenting**, where a group of three or more individuals become genetic parents of a child. The technology that one day could make solo parenting possible is based on Shinya Yamanaka's 2006 discovery that cells from adults could be turned into stem cells: "By introducing only a few genes, he could reprogram mature cells to become pluripotent stem cells, i.e. immature cells that are able to develop into all types of cells in the body" ("The Nobel Prize"). For his discovery, Yamanaka shared a Nobel Prize in Physiology or Medicine in 2012.

So far, experiments in mice confirm that it is possible to produce sperm and egg cells from any cells of an adult (Palacios-González et al. 2) Theoretically, nothing prevents a procedure in which both the sperm and the egg would come from the same individual. Sonia Suter suggests that for single individuals it would be possible to derive sperm or egg cells "that would complement their naturally produced" sperm or egg cells. At the same time, Palacios-González and his coauthors point to a problem which might result. The lack of genetic diversity would make children born of solo parents even more susceptible to genetic disorders than the children of "siblings or first-degree cousins" (5).

Despite using only the genetic material of one individual, solo parenting is not the same as cloning. As Daniela Cutas and Anna Smajdor say in their 2016 article, "Solo reproduction would be akin to reproducing with one's own identical twin." As Palacios-González and his coauthors put it, "This self-breeding setting would be different from reproductive cloning" (5). While cloning amounts to producing a child who is a genetic twin of the person cloned, solo parenting creates a baby who has, to put it in simple terms, all the same component pieces as the solo parent, but arranged in a different order.

Such a child's relationship to its progenitor has no legal definition. In fact, Suter argues that a child created through solo parenting technology "would have no genetic parent" because according to the current legal definition of parentage, a parent is a person who contributes exactly 50 percent of the genome (22). A solo parent would contribute 100 percent.

While the technology that would make solo parenting an option is still in the developmental stages and has not been tried in humans, SF writers have thought about the idea. Robert Heinlein constructed a story titled "All You Zombies—" where a time traveler manages to be his own father and mother. Unfortunately, in the light of the technology that we will soon have, Heinlein got it wrong. His narrator, who is father, mother, and baby is also clearly the same genetic person in each of those incarnations. In effect, the result of this adventure in solo parenting appears to be a clone, and that would not be the actual outcome.

Still, SF writers can't be expected to be right all the time. Sometimes it's enough to raise the subject and suggest or reinforce a new idea, however strange it may seem at first. Even when it is more muddle than example or to put it another way, more mess than quest, it can be illuminating.

Stepford Wives

I once suggested in print that the term comes from a story "where women are replaced by docile android replicas with limited vocabularies and insatiable desires to clean house" ("The Taming of the Shrew"). Whether they are androids or **robot**s is debatable, and the definition is a bit more complicated as well. It comes from Ira Levin's 1972 novel (and the 1975 cult film based on it) *The Stepford Wives*. In both, the men of the Stepford, Connecticut, Men's Association replace their wives with obedient robots who are younger, sexier, more sexually active and better housekeepers than the human women were.

The term has come to mean wives (or anyone else) who subordinate their own interests and individuality to another person or group or who look and act so as to fit a standard pattern. The term (in the form of the adjective "Stepford") has been applied to husbands, beauty queens, models, and even to students who are more interested in conforming to rules and staying out of trouble than asserting what some might see as their rights or freedoms.

As she is trying to figure out what's going on in Stepford, Joanna Eberhart tells her therapist, "If only you could see what Stepford women are like. They're actresses in TV commercials, all of them. No, they're not even that." She continues to struggle and finally suggests that the Stepford wives are like Disney animatronics (112). In the 1975 film, her conversation is considerably more urgent, and she tells her therapist that she is about to be replaced by something that will look the way she does, but she says, "She won't be me. She'll be like one of those robots in Disneyland." The film makes it clear that one of the robot replica's first tasks is to kill the human it's about to replace (in this case with a nylon stocking).

There were three television movie sequels in which the conformity and docility were imposed by chemical means. They were *Revenge of the Stepford Wives* (1980), *The Stepford Children* (1987), and *The Stepford Husbands* (1996). In 2004, there was an unsuccessful comic remake of the original film that removed most of the feminist issues.

The robots in the novel and original film do not appear to be sentient. (See **Artificial Intelligence** and **Deep Blue**.) The activities they pursue are limited, they are unable to deal with emergencies or even unfamiliar events, and their vocabularies are limited to the words their human originals were tricked into recording. In addition, the emphasis on Disney animatronics seems to be designed to create a sense of unease in movie audiences, the same kind of sensation that robots and animated figures are supposed to produce when they look almost human. (See **Uncanny Valley**.)

Teleportation

Moving an object or person from one place to another in a single leap with no intervening form of transportation. The idea of such instantaneous travel exists in early religious narratives. Though the OED doesn't find the word itself until 1931, as early as 1877, Edward Page Mitchell was writing about what he called a "Telepomp," a machine, modeled on the telegraph, with which one of his characters could disintegrate matter, pass it through a wire and rebuild it "at the receiving instrument" ("The Man without a Body").

There are two main kinds of teleportation in science fiction. One is a psychic power as in James Blish's *Jack of Eagles* (1952), Alfred Bester's *Tiger! Tiger!* (1956, also published in 1957 as *The Stars My Destination*), Robert Heinlein's *Stranger in a Strange Land* (1961), Roger Zelazny's *Creatures of Light and Dark-*

ness (1969) and, more recently, Timothy Zahn's *A Coming of Age* (1984) and Steven Gould's novel *Jumper* (1992), plus the film made from it (2008).

More often, the word means matter transmission, or converting solid objects (including humans) into energy and then reconstituting them at the other end of their journey as in Poul Anderson's *The Enemy Stars* (1959). Clifford D. Simak uses a modified version of such a system (transporting patterns only) in *Way Station*, and, of course, in *Star Trek*, Scotty (via the transporter) is constantly beaming Captain Kirk from one dangerous place to another.

The *Enterprise*'s method for beaming people up (and down) is undoubtedly the most famous teleporter in history and one of the main prompters for discussion of and research into teleportation (though *Stargate*'s Ring Transporter runs it a close second). It was "invented" by Gene Roddenberry as an alternative to landing his massive space vessel on a different planet every week, and it provided an immense savings for his special effects budget. In terms of current or near-future technology, it is, however, not much removed from the near magic of esper powers.

Physicist Lawrence M. Krauss, in his *The Physics of Star Trek* (1995), estimated that with the computer storage facilities then available, the information necessary to rebuild a human being would require (figuring 10 gigabytes per 10 cm disk) a stack of disks that "would reach a third of the way to the center of the galaxy—about 10,000 light years, or about 5 years' travel in the *Enterprise* at warp 9!" (77).

Nor is the time to process such information negligible. Krauss suggested that at the then current top digital transfer speed of 100 megabytes per second, "It would take about 2000 times the present age of the universe ... to write the data describing a human pattern to tape!" (77).

In addition, the resolution necessary for accurate scanning of targets on a planet's surface (since as a rule the *Enterprise* is not moving humans from one transporter pad to another) "would need a telescope with a lens greater than approximately 50,000 kilometers in diameter!" (Krauss 83). Although Krauss seems inordinately fond of exclamation points, given the size of the difficulties he was describing, his excitement may have been justified.

Still, information storage and transmission should be a soluble problem, and Krauss was convinced that if computer innovations continued at their then current rate, for the *Enterprise*, "This is one area that could possibly be up to snuff in the twenty-third century" (78). Or earlier, of course, with alien technology and transporting only patterns, as Simak does in *Way Station*. However, depending on the method employed, computers may well be the easy bit (yes, I know), and we are far from reaching the end of the problem (or this entry).

The *Star Trek: The Next Generation Technical Manual* describes a part of the beaming process with this abominable example of *Trek* tech-speak, "The molecular imaging scanners derive a real-time quantum-resolution pattern image of the transport subject while the primary energizing coils and the phase transition coils convert the subject into a subatomically debonded matter stream" (Okuda and Sternbach 103).

Or, in English, a person is scanned, melted down to the quark level, and beamed out. That means sending not only information but also matter, and I say melted down because Krauss maintains that the only way to "debond" matter to the quark level using a finite quantity of energy is "to heat up the nuclei to about 1000 billion degrees (about a million times hotter than the temperature at the core of the Sun)" (73).

Poul Anderson uses a system something like this in *The Enemy Stars* (1958). There, scanning a person on the subatomic level reduces her (or him) to superheated plasma, which (unlike the *Star Trek* version) is stored to provide material for the next incoming pattern. But even if teleportation were technically feasible, there is a serious possible problem with this technology, as Larry Niven pointed out in "Exercise in Speculation: The Theory and Practice of Teleportation" (*All the Myriad Ways* 88–89). If the machine simply picks up a person at point A and moves him to point B like a sort of subatomic bus, all is well and good. We have not, so far, looked at any such system, but Niven suggests the tunnel diode effect in quantum physics may point the way (88).

As soon, however, as we start reducing the passenger to atoms and patterns and recreating him either from the same or (and this is much worse even if it is technically much easier) different solid material, we have troubles and we may have murders. Is the passenger who emerges and is reconstituted merely a copy of the original, the original having been killed in transport? And how would anyone tell? The answer is that only someone who had been through the experience and emerged alive and whole would know that it was safe, but, unfortunately, all the copies of murdered originals, because they share the same memories, would believe that they had safely traversed the distance and so are unreliable witnesses.

In other words, you have to try it yourself to be sure—or dead—and even then what you think you remember may be suspect. In fact, only a memory of actually existing in the strange world of the transporter beam à la Lieutenant Reginald Barclay in the *STNG* episode "Realm of Fear" would be truly reassuring since the copy of a murdered original would "remember" stepping into the transporter and stepping out but not the time (and death) in between. Barclay's experience is, of course, problematic in itself. As Phil Farrand asks in his *Nitpicker's Guide for Next Generation Trekkers,* "Doesn't it seem as if Barclay is staying conscious for a *long* time through this process of molecular *deconstruction?*" (382).

One way to solve the problem Niven identified would be for the person being transported to remain conscious throughout and to experience a transfer of consciousness from one location to the other, perhaps with a brief interval of seeming to be in both places at once. I suspect, though, that this would prove nearly impossible to achieve except in a system like Simak's, where the individual moves from body to body.

"Second Chances," another *STNG* episode, suggests that transporters on the *Enterprise* can be weirdly unreliable, in this case creating another Riker whose existence isn't discovered for eight years. The Riker on the *Enterprise* is naturally concerned about his status now that he has a transporter-created twin. He asks,

"Which of them is real?" Geordi La Forge (who apparently hasn't read his ship's technical manual or doesn't believe it if he has) answers, "Well, that's the thing—both. You were both materialized from a complete pattern."

According to La Forge's statement, the *Enterprise* does not convert matter into energy and send it from place to place; instead, it transmits patterns, and two Rikers can be equally genuine if both came from a complete pattern. This is even worse than the situation in the Original *Star Trek* episode "The Enemy Within," where Captain Kirk is split into his good and evil halves. That incident also suggests pattern and not matter transfer (How could we get two full-size Kirks from the mass required to make one?), but it argues for a transfer of something else—life force, consciousness, or soul—as Simak also seems to do. Therefore, two Kirks made from one will be incomplete human beings, having only half of a personality or whatever we wish to call it, and unable ultimately to survive apart.

The Rikers, on the other hand, appear to be mere copies of an original which has probably been destroyed, each equally valid or invalid. In *Teleportation: The Impossible Leap*, David Darling discusses the possibility that once a transporter has produced its copy, the original, should she or he have survived through some glitch in the system, might have to be eliminated. He writes, "Killing the original may then be considered simply a way of disposing of an unnecessary and troublesome copy" (250).

At issue here is the very nature of being. Krauss, who clearly hasn't talked to many religious fanatics (or science fiction fanatics either), announces that successfully recreating a person from the quarks up "flies in the face of a great deal of spiritual belief about the existence of a 'soul' that is somehow distinct from one's body" (69). Simak, however, seems to be counting on something like a soul or at least a consciousness to make his system work. He says, "The materializer had built up a pattern of it—not only of its body, but of its very vital force, the thing that gave it life" (62). Thus, although there is "a long trail of dead," the individual goes "on and on" (63). The dead bodies left behind in the materializer tanks are destroyed by chemicals, but clearly the aliens are not killed by the station operators or by any active part of the transport process but are left as mere dead matter when the life force has moved on.

In his nonfiction book *The Physics of Immortality*, Frank J. Tipler argues that a sufficiently exact computer copy of a dead human being will amount to a resurrection. He says, "The quantum criterion for system identity applies to humans, and hence two humans in the same quantum state *are* the same person: if a 'replica' of a long-dead person were made which was identical to the quantum state of the long-dead person, the 'replica' would *be* that person" (234).

Tipler is making a case for computer-generated resurrection and immortality (something that Spider Robinson also does in his 1997 novel *Lifehouse*, 200–201), but his argument applies equally well to transporter metaphysics. Incidentally, if he is right, there could be only one true Kirk (or even Riker) because (Tipler maintains) no two humans could exist simultaneously in the same quantum state.

I don't think, though, that the possibility of similar Rikers in similar quantum states is precluded. Jerry Oltion's 1997 *Star Trek* novel, *Mudd in Your Eye*, actually postulates an alien civilization which resurrects its citizens by scanning them at frequent intervals and recreating those who happen to die violently by using the last recorded pattern when they were "still alive and in good health" (228–29). Several *Enterprise* crewmembers (including Kirk) are vaporized and then "resurrected" from their patterns, and oddly enough no one (not even Spock) asks any awkward questions about the procedure. David Darling suggests such a system in all seriousness, "Then if anything happened to the living you, the backup could be used as a replacement" (250). (See **Neural Lace**.)

However we look at them, transporters (including *Stargate SG-1*'s Stargate, which seems to operate in similar fashion) are clearly perilous pieces of technology, full of far more uncertainties than even Heisenberg dreamed, so perhaps Larry Niven should have the last word on the subject. He says, "I wouldn't ride in one of the goddamned things" (89). The issue may not be entirely hypothetical because, as usual, science fiction quickly turns to science fact. In June of 2004, Rainer Blatt and David Wineland published independent papers that "described using laser pulses to transfer information from one atom to another in a different location" (Svitil 72). And in May of 2014, physicists "at the Kavli Institute of Nanoscience in the Netherlands published a paper in the journal *Science* reporting their ability to teleport information three meters" (Curry). The distance is greater than in earlier studies, and "the results ... have an unprecedented replication rate of 100 percent" (Statt).

These are very small first steps (200 micrometers in the first instance and about ten feet in the second), but it is very important as a demonstration of "entanglement," a quantum phenomenon "where particles ... share information even if they are physically separated" (Svitil 72) and as a stage on the way to making a transporter or at the very least, a quantum computer.

Time Bunny

A pretty girl or an attractive male from an alternate (alternative) or other timeline who hangs around with time travelers. It was most likely formed after ski bunny or snow bunny. The *OED* defines snow bunny as (second meaning) "a pretty girl who frequents ski slopes." "Ski bunny" can apply to either sex, as in Bunny slope, a place for beginners to practice skiing. The term was originated by Bruce Sterling and Lewis Shiner in their 1985 story "Mozart in Mirrorshades." In an alternate timeline, Mozart says, "You can bring your little time-bunny along if you want."

Rice, the character Mozart is talking to, is worried about something more than the trip from Paris to Salzburg. His concern is what will happen if "the project" is shut down. "He wanted to take her with him" into "**Realtime**." "He wanted to do the decent thing, not leave her behind in a world without Hersheys and *Vogues*." The woman in question is Marie Antoinette. The term is an example of the trivialization of the past and the destruction of individuality and culture by multi-national corporations, a frequent theme in cyberpunk stories.

Oddly enough, no one in the *Doctor Who* series seems to employ it. As the Mozart character uses it, the term is clearly sexist, and that is part of the condemnation of the story's Realtime society. However, as my definition makes clear, there are male time bunnies as well. Sutherland, one of the women from the future, as Rice puts it, "goes for the locals." Young Mozart, "who might almost have passed for a teenager from Rice's time," falls within the parameters of the definition as he desperately attempts to get a Green Card that will allow him to emigrate to Realtime, where he's "number five on the *Billboard* charts."

Time Machine

Coined by H. G. Wells in his novel *The Time Machine* (serialized in *National Observer* 1894, book publication 1895), it means a device for traveling forward or backward in time. The "Time Traveller" (as Wells always calls him) describes the process when he introduces a smaller model of the mechanism to his friends, "This lever, being pressed over, sends the machine gliding into the future, and this other reverses the motion. This saddle represents the seat of a time traveller. Presently I am going to press the lever, and off the machine will go. It will vanish, pass into future time, and disappear" (Wells 40).

Wells was by no means the first person to think of using a passage through time as a literary device. (See **Cryonics**.) There had been many other journeys, but they had not, with one exception, been made possible by a mechanism. That exception was Edward Page Mitchell's "The Clock That Went Backward," published in 1881. Though it's a minor work, it contains interesting Hegelian speculations and a clock whose "hands were whirling around the dial from right to left with inconceivable rapidity. In this whirl we ourselves seemed to be borne along. Eternities seemed to contract into minutes while lifetimes were thrown off at every tick." It too is a machine for traveling in time.

Wells' novel was a breakthrough event. The many fictional explorations of the nature of time that have spanned possibilities from the scientifically plausible to the plainly impossible have grown, if not from the seed he planted, at least in the shade of the flourishing tree of his imagination. I do not mean to suggest that Wells intended his time machine as anything like a serious speculation, but only that his "magic trick," or means to a narrative end, set the stage for more serious performers.

In a discussion of "the whole business of the fantasy writer," Wells declared, "In my student days we were much exercised by talk about a possible fourth dimension of space; the fairly obvious idea that events could be presented in a rigid four dimensional space time framework had occurred to me, and this is used as the magic trick for a glimpse of the future that ran counter to the placid assumption of that time that Evolution was a pro-human force" (Wells "Preface" iv–v).

Not surprisingly at the beginning of his career, Wells' main interest was in social commentary and criticism rather than scientific exploration and prediction. He cites Daniel Defoe as his inspiration and declares, "These tales have been compared with the work of Jules Verne.... As a matter of fact there is no literary resemblance whatever between the anticipatory inventions of the great Frenchman and these fantasies" (iii).

However, Wells may be underestimating his own achievement. Paul Davies writes, "I first read.... *The Time Machine* as a teenager, and it left a lasting impression on me. In fact, it probably contributed to my determination to become a scientist" (*About Time* 233). Such an impact is not uncommon for science fiction in general or for works that involve time travel in particular. Theoretical physicist Kip Thorne says, "Without Carl Sagan's phone call and the challenge to make his novel [*Contact*] scientifically correct, I would never have ventured into research on wormholes and time machines" (500). Davies also says about the 1895 publication of Wells' book, "Although this was a full decade before the special theory of relativity, Wells anticipated some aspects of Einstein's time with uncanny accuracy."

The Time Machine shows us what has been called block time or the block universe, which Wells had previously considered in 1888 in *The Chronic Argonauts* (Wells [Stover, ed.] 28, 174). That is one of the possible consequences of visualizing time as a fourth dimension. In the words of Paul Nahin, "The view of reality that the past and present and future are joined together into a four-dimensional entity called *spacetime* is due to the work of Hermann Minkowski.... Einstein's mathematics professor when he was a student at Zurich" (101). Nahin argues that in 1908, Minkowski not only demonstrated the nature of time in Einstein's universe, he also "gave mathematical expression to the philosophical exposition of Wells' Time Traveller" (102).

Indeed, here is how the "Time Traveller" puts it, "I have been at work upon this geometry of Four Dimensions for some time.... For instance, here is a portrait of a man at eight years old, another at fifteen, another at seventeen, another at twenty-three, and so on. All these are evidently sections, as it were, Three-Dimensional representations of his Four-Dimensioned being, which is a fixed and unalterable thing" (Wells 30).

As the words "fixed and unalterable" indicate, the notion of the block universe (also sometimes called eternalism) suggests that past, present and future all exist simultaneously. One of the clearest expressions of the idea is in the 1933 film *Berkeley Square*. It is based on John L. Balderston's play *Berkeley Square*, which premiered in London in 1926, and behind that on an unfinished novel by Henry James titled *The Sense of the Past*.

Peter Standish, who takes advantage of a **timeslip** to travel into the past, says, "Suppose you're in a boat sailing down a winding stream. You watch the banks as they pass you. You went by a grove of maple trees, upstream. But you can't see them now, so you saw them in the past, didn't you? Now, you're watching a field of clover, before your eyes, this moment, in the present. But you don't know yet what's waiting for you round the bend in the stream ahead; there may be wonderful things, but you can't see them till you get round the bend, in the future, can you?"

So far, it seems quite normal, except for the little uneasiness caused by Standish laying out past, present and future in a landscape, but he continues, "All right ... remember, now, you're in the boat. But I'm up in the sky above you, in a plane. I'm looking down on it all. I can see it all at once. So that the past, the present, and the future of the man in the boat are all one to the man in the plane. But doesn't that prove that all Time must really be one? Time—real Time—is nothing but an idea in the mind of God!" And just in case there's any doubt, Peter Standish says of the ancestor who shares his name, "I think they're alive. Peter Standish is alive. Just as I'm living here, he's living there—back in his own time."

Apart from the divine observer, *The Mothman Prophecies* (2002) makes the same point. Appropriately in this case, former physics professor Alexander Leek does the explaining of why the film's mysterious aliens seem to be able to see into the future, "Hey, look up there. If there was a car crash ten blocks away, that window washer up there could probably see it. Now, that doesn't mean he's God or even smarter than we are, but from where he's sitting, he can see a little further down the road."

Interstellar, with Kip Thorne as scientific advisor and executive producer, is considerably more complex than the films I've been quoting, but what Dr. Amelia Brand says about the "Bulk Beings" in *Interstellar* (2014), sounds very familiar: "They are beings of five dimensions. To them time might be another physical dimension. To them the past might be a canyon that they can climb into and the future a mountain they can climb up, but to us it's not."

Block time (or the block universe) not only explains what H. G. Wells had in mind for his time machine, it also explains the enigmatic response of the character Kosh Naranek on *Babylon 5*. In the 1998 television movie *Babylon 5: In the Beginning*, Ambassador Delenn asks, "Are you there?" and Kosh replies (as he does in other situations), "We have always been here." In short, the past is alive, the future is already in place, and if only a time machine could be built, there are plenty of places to go.

Timequake

A disturbance similar to an earthquake but caused by a time paradox or other substantial alteration in the timeline, capable of disturbing or destroying persons, events, or if the timequake is severe enough, the timeline itself. Murray Leinster suggested something like it, though for different reasons, in 1934 in "Sidewise in Time," "There has been an upheaval of nature, which still continues. But instead of a shaking and jumbling of earth and rocks, there has been a shaking and jumbling of space and time" (289). However, the term "timequake" comes from the 1989 film *Millennium*, which concerns time travelers who snatch airline passengers just before they die and transport them from the past into the future. The underlying assumption is that such actions will have minimal impact on the timeline (and healthy people are desperately needed in the sick and polluted future), but, of course, things go wrong.

Coventry, who is in charge of the operation, is told, "Sir, I have a timequake approaching." This is followed by a voice over a loudspeaker: "Paradox! Timequake approaching, force 3!" Coventry blames one of his agents, "A paradox, Louise, you've changed the past. I know damn well we can't change the past! It catches up with us. We change." He continues after Louise's objections and justifications, "One not much bigger than that could destroy us completely." John Varley based the screenplay for the film on his 1977 short story "Air Raid," but the term does not appear there.

The *Star Trek* franchise has a similar idea with a cluster of similar terms, "temporal shock wave," "space time shockwave," and "temporal wavefront." In the *Star Trek Voyager* two-part episode "Year of Hell," Tuvok says, "It appears to be a massive build-up of temporal energy. Some kind of space-time shock wave." Captain Janeway says, "Display the scans we made of this region before the temporal shockwave hit." According to *Memory Alpha*, the definition for any or all of the three terms is "a powerful buildup of temporal energy which travels through space, changing history as it propagates" ("Temporal Shock Wave"). For the term timequake or any of the others to be applicable, time travel to the past must be possible, and beyond that, it must be possible for time travelers to change their own past, two assumptions that are currently under dispute. The ability to travel into the future as well as the past is often added to these scenarios, as it is in *Millennium*, but it is not clear that traveling in one direction makes the possibility of traveling in the other direction (e.g., returning to when one began) likely. Of course a **timequake** could be created only by changing the past. See **Realtime**.

Timeskip

The title of a story by Charles de Lint. Jilly, the character within the "ghost" story who indicates she coined the word, says, "It's like a broken record.... It just keeps playing over and over again, only unlike the record it needs something specific to cue it in" (93). In this instance, the "something specific" is rain. Jilly presents a timeskip as a form of **timeslip**, which she defines as "a bit of the past slipping into the present" (93). The story looks on the surface very much like a standard ghost story, but it turns out to be time travel instead. A man walks up and down a particular street every time it rains, and then vanishes suddenly, as though he were "a restless spirit with unfinished business" (93). Instead, he is trapped in a timeskip. The man is freed by the narrator's girlfriend, who is pulled into the past and trapped there, an event later confirmed by a photograph of her from 1912. The narrator too is pulled into the past, but he manages to escape with help from Jilly, who had warned of the dangers in advance, "Sometimes a ghost like that can drag you back to whenever it is that he's from and you'll be trapped in his time. Or you might end up taking his place in the timeskip" (94).

The story is a particularly good example of **speculative history**, which, most often, is another name for science fiction, but in this case, is a remarkably eerie horror story with a scientific explanation. As a bonus there's something that is very common in SF, the clear feeling that the movement of time from past to present with its attendant progress has resulted in a "now" that is considerably better than "then." Beyond the separation of the lovers, being trapped in the past seems to be a terrible thing in and of itself.

Timeslip

A part of the past or future which somehow slips into the present or a part of the future or the present which somehow slips into the past. (Compare **Timeskip**.) In other words, a mysterious form of time travel without a time machine or other mechanism to make it possible.

The result, for the writers at least, is a considerable simplification of explanation and exposition. How did it happen?—Who knows? Is time travel possible?—Well, yes, in this case it was.

To take two relatively recent examples, Diana Gabaldon's *Outlander* (the novel, its sequels, and the television series) in spite of being initially inspired by *Doctor Who*, has an easier "time" of it than the 2016 television series *Timeless* (Gabaldon). Touching a monumental boulder is considerably simpler to describe than constructing a time machine that can deal with "a closed timelike curve" and "create a powerful enough gravitational field" so that it "could actually bend it [the curve] back on itself, creating a kind of loop that would allow you to cross over to an earlier point." Not to mention, on film at least, building something that looks like a time machine and adding the special effects.

The *OED* cites L. Sprague de Camp in *Lest Darkness Fall* as the first to use the term time-slip (in 1941). Jeff Prucher's definition goes some way to helping writers with their explanations. It is, he says, "a rift or flaw in the fabric of time that allows travel between two or more periods of time or timelines" (242). As almost always happens, the idea predates the term even though the magazine publication of *Lest Darkness Fall* came in 1939.

Mark Twain's *A Connecticut Yankee in King Arthur's Court*, which has more than a few similarities with L. Sprague de Camp's novel, came in 1889. While being hit in the head "during a misunderstanding conducted with crowbars" (5) seems to have more in common with sleep or dreams as time travel than with walking through a rift or portal, Twain's narrator suggests that what happened should be called "transposition of epochs" (1), which could easily serve as a definition for timeslip. John L. Balderston's play *Berkeley Square*, which premiered in London in 1926, opened on Broadway in 1929, and was filmed in 1933, presents a very clear example of a timeslip, though it does not use the word. (See **Time Machine**.) The shift between time periods is effected by passing through a door.

Another clear use of the idea of a timeslip came with Alison Uttley's 1939 novel *A Traveller in Time*. For that story's narrator, Penelope Taberner Cameron, the transition was more difficult, though not entirely different, "I flung open the door, and I fell headlong down a flight of stairs. I had dropped into the corridor where I had seen the servants pass with their jugs and tankards. For some time I lay half-stunned with surprise, but unhurt, for I had fallen silently like a feather floating to the floor" (68).

As with other forms of time travel, sometimes the people who make use of timeslips are able to change things and sometimes they are not, depending on the writer's interpretation of the laws of physics. Briefly, the issues are that travel to the past makes three relatively plausible outcomes possible. The first (and, most people think, the most likely) possibility is that the past is fixed and everything that can happen has already happened. By that argument, time travelers to the past will do what they have already done, and the past will not be changed. The second possibility is that the past can be changed, and the future from which the time traveler came will be altered or even destroyed by those changes. (See **Timequake**.) The third possibility is that time travelers can make changes, but those changes will not affect the timeline from which they came. Instead, the changes will create an alternate (alternative) or parallel timeline in which the time travelers may or may not be trapped.

In *Lest Darkness Fall*, for example, Professor Tancredi declares that though time travelers can "change all subsequent history," there is no possibility of a paradox. As he puts it, "The trunk of the tree of time ... continues to exist. But a new branch starts out where they come to rest" (1). In *A Traveller in Time*, it is harder to shift events. The attempt to save Mary Queen of Scots fails, and Dame Cicely exclaims, "It was not to be. Poor lady, she is doomed" (Uttley 311).

S. M. Stirling's *Island in the Sea of Time* (and its sequels), moves Nantucket Island three thousand years into the past and creates an **alternate history**. The 2001 film *Kate and Leopold* involves travel through a temporal portal or rift with the end result that Kate McKay starts in the present and ends in 1876, where she was always meant to be since that was what had already happened. This is the notion that the past cannot be changed by travelers because anything they do is what they have already done—the past is fixed.

The idea of the timeslip is one of many examples of the nearly universal fascination with history, with time and with the future that a judicial mixture of the two will create. See **Bobble, Chronotransference, Cryonics, Realtime, Speculative History,** and **Time Bunny**.

Uncanny Valley

The more closely nonhuman things such as **robot**s, computer programs and animated characters resemble human beings, the more affection and affinity we feel for them, until they are nearly identical with humans but not quite. Then, they inspire a sense of uneasiness or even dread in some people. This strange, near-human land is what **robotics** professor Masahiro Mori called the uncanny valley in an article in *Energy* in 1970.

The term, which was first used in English in 1978 by Jasia Reichardt, has become part of the ongoing debate about the place of robots and androids in human society. It has also been used in discussions of animation and software. In *Robots: Fact, Fiction and Prediction*, Reichardt presented a translated version of Mori's article, graphs and all (26–27). In what amounted to a thought experiment, Mori defined his place of deep creepiness as lying between the peaks of a "stuffed animal" on one side and a "healthy human" on the other, and his recommendation was, "We hope to design robots or prosthetic hands that will not fall into the uncanny valley." There is much argument about what the uncanny valley is, what generates the feelings of unease and, indeed, whether the phenomenon is real.

Mori's concept, as is usual with definitions of the uncanny itself, seems to have more to do with his own feelings than with reality. He wrote, "Since negative effects of movement are apparent even with a prosthetic hand, a whole robot would magnify the creepiness. And that is just one robot. Imagine a craftsman being awakened suddenly in the dead of night. He searches downstairs for something among a crowd of mannequins in his workshop. If the mannequins started to move, it would be like a horror story."

I don't understand his attitude toward prosthetic hands and the example he gives of mannequins suddenly beginning to move is not relevant, however much it might resemble an episode of *Doctor Who*. Of course, it would be strange if mannequins suddenly started to move. It would also be strange if the furniture began to dance in the dark, but that should not affect the future design of wheel chairs.

From the beginning of serious attempts to discuss what has been called the "uncanny" (apart from anything to do with robot valleys), there has been disagreement. The two most important early explorations are by Jentsch and Freud. In fact, Frank Pollick suggests that in Mori's original Japanese combination "bukimi no tani," "'bukimi' has several translations including 'eery,' 'strange' and 'uncanny'" but "'uncanny' was chosen due to its psychological resonances" with Freud's essay and its translation into English (71).

Ernst Jentsch in his 1906 essay "On the Psychology of the Uncanny," says he will not "attempt ... to define the essence of the uncanny" because "the same impression does not necessarily exert an uncanny effect on everybody" or even on the same person every time (1–2). For Jentsch, it is the new and unfamiliar that causes the sensation, and he too points to automata and to the literary technique of creating the uncanny by leaving "the reader in uncertainty as to whether he has a human person or rather an automaton before him" (11). He gives E. T. A. Hoffmann as an example of a writer who has repeatedly done this.

In his 1919 essay "The Uncanny," Sigmund Freud did a thorough job of proving that the truly uncanny element in Hoffmann's story "The Sandman" (one of the sources for Offenbach's opera *The Tales of Hoffmann*) was not the doll, Olympia, or any doubts about her reality, which "the author ... treats ... with a faint touch of satire," using "it to make fun of the young man's idealization of his mistress" (5). Instead, Freud quite rightly identifies the Sandman's stealing of eyes as the true horror.

Freud's most interesting contribution to the controversy may well be his identification of "the idea of a 'double' in every shape and degree" (9) with the uncanny in Hoffmann's stories and elsewhere. (See **Alter Ego**.) Sara M. Watson, for example, finds an uncanny sense of the double in what others see as her online persona, "Data tracking and personalized advertising is often described as 'creepy.' Personalized ads and experiences are supposed to reflect individuals, so when these systems miss their mark, they can interfere with a person's sense of self. It's hard to tell whether the algorithm doesn't know us at all, or if it actually knows us better than we know ourselves." (See **Internet of Things**.)

Clearly, Mori's thought experiment is not the only one to deal with the future of artificial people, which, if Jentsch and Freud are right, is a logical place to find the uncanny. Here are just a few examples. Isaac Asimov introduced Daneel, his most extraordinary humaniform robot, in 1953 in *Caves of Steel*. It was a story of struggles and transformations, of a society coming to terms with robots, and of a bigot overcoming his prejudice against them. In 1961, William Welch's play *How to Make a Man* opened on Broadway. It was based on "How-2" by Clifford Simak, and the center of the story was a legal case about the rights of robots ("How to Make a Man").

In 1982, the film *Blade Runner* created "replicants," android slaves who were nearly human or perhaps indistinguishable from humans or possibly better than humans, depending on the various interpretations (and versions) of the movie. In 1983, in the last volume of Asimov's "Daneel" trilogy, former bigot Lije Baley thinks, "Why dismiss them with a word—machines? They were *good* machines in a Universe of sometimes-evil people. I have no right to favor the machines vs. people subcategorization over the good vs. evil one. And Daneel, at least, I cannot think of as a machine" (223).

In *Foundation's Triumph* (Book 3 of the Second Foundation trilogy), David Brin goes even further. A computer recreation of Joan of Arc says, "*After getting to know Daneel Olivaw, I came to recognize a true apostle of chaste goodness and saintly self-sacrifice. His followers wield righteousness, for the sake of countless suffering human souls*" (18; italics in original).

Writing in the January 9, 2015, *New York Times Magazine*, Robin Marantz Henig suggests that such very positive views of robots may turn out to be correct, "The experts tend to be optimistic about robots' ethical prospects. Wallach talks of a 'moral Turing test' in which a robot's behavior will someday be indistinguishable from a human's. Scheutz goes even further, saying that one day robots will be even more morally consistent than humans."

In 1992, in his novel *Cold as Ice*, Charles Sheffield included a sentient "fax" named Mord, a version of a human named Mordecai Perlman, "stripped of all material attributes" and left with nothing but "the joys of pure intellect" (204). In its seven-year run between 1995 and 2001, *Star Trek: Voyager* explored a cyborg reclaiming her humanity and a hologram achieving sentience. By 2002, a science fiction romantic comedy film called *Simone*, presented an intelligent computer program successfully masquerading as a movie star.

And in 2015, the San Diego Repertory did a play by Thomas Gibbons where a man's personality is downloaded into a robot which is first assembled while the audience watches. The title of that work, not surprisingly, is *The Uncanny Valley*. Also in 2015, the British-American television series *Humans* began airing on AMC. Based on a 2012 Swedish series called *Real Humans*, it presents a society where nonsentient synthetic humans have taken over many human tasks in the workplace and the home.

It may now be time to ask if any of these ideas or images or even the robots themselves are new and strange enough to be creepy. Science fiction has made them all familiar, just as science is working to make them all real. Erik Sofge writes in "The Truth About Robots and the Uncanny Valley," "The uncanny effect appears to be an incredibly specific and specialized phenomenon: It seems to happen, when it does, remotely. In person, the uncanny vanishes. There's nothing in the way of peer-reviewed evidence to support this, but then, there's almost nothing to confirm the uncanny effect's existence in the first place. As an unsupported theory that has morphed into a nerdy breed of urban legend, anecdotes are all we have to work with."

One frequently cited example of how not to fall into the uncanny valley is James Cameron's *Avatar*, where the motion-capture aliens are large and blue. "But the filmmakers also created a digital version of Jake Sully—Worthington's character—in his human form, allowing them to integrate Jake into scenes set in the fully digital realm of Pandora" (Talbot). The human Jake Sully also avoided the valley. Indeed, experiments indicate that the most likely cause of uneasiness is when something looks human but fails to move as humans move (Kiderra).

Beyond that, as Jentsch understood, the experience of the uncanny is inconsistent from person to person and even from time to time for the same person or for an entire culture. The study that demonstrated the connection between the uncanny and nonhuman movement selected twenty test subjects who "had no experience working with robots and hadn't spent time in Japan" (Kiderra). The feeling can still be generated by clowns, dolls, humans made up to look like *anime* characters, or just plain incompetent actors. For this last category, it would be hard to improve on Hamlet's words, They "have so strutted and bellowed that I have thought some of nature's journeymen had made men and not made them well, they imitated humanity so abominably" (Shakespeare 3.2. 29–31).

For most roboticists, programmers and movie makers, the goal is to avoid that fault altogether, making their artificial people indistinguishable from the real thing. When will that happen? On film it is happening now. For robots it is coming soon: "When one roboticist named Peter Kahn visited Karl MacDorman's human-computer interaction lab at Indiana University and wanted to take apart Ishiguro's Repliee Q1 Expo, a petite Japanese humanoid-woman in a pink blazer, he first turned to his wife and asked, 'May I touch her?'" (Eveleth).

To quote the sales pitch of John Woods, the roboticist-entrepreneur on the CBS television series *Extant*, "It turns out the true Uncanny Valley isn't visual at all; it's the value of genuine connection. The goal of the Humanichs Project is to bridge that divide by bringing humanity to the machine" ("Re-Entry"). As so often happens when SF and science interact, that goal also exists in the minds of real scientists working in actual laboratories.

Uplift

The term comes from David Brin's 1980 novel *Sundiver*, and like the idea itself, still has some areas of disagreement. Prucher's *Brave New Words* defines it as "the making of a sentient species from a non-sentient one" (255). While that is both succinct and clear, it leaves open the ever-more vexed question of which species in addition to humans qualify as sentient already, and for those that do not, what might be required to raise them to that level.

Brin's definition in *Startide Rising* (sequel to *Sundiver*) neatly avoids such problems, "The process by which older spacefaring races bring new species into Galactic culture, through breeding and genetic engineering" (xiv). But alas, the standards of Galactic culture or even a general sense of what constitutes a spacefaring race is not yet available.

The Encyclopedia of Science Fiction defines Uplift as "an assisted leap of Evolution—specifically, the raising of nonsentient or otherwise handicapped beings to a level of Intelligence or technological capability comparable to or exceeding humanity's" ("Uplift"). But again, the notion of "comparable to" may be difficult to pin down, and "otherwise handicapped" is not much better. It might mean lacking the type of appendages or the kind of environment that leads on to the machine culture which humans see as civilized, but there are many other possibilities.

The idea of Uplift is considerably older than the term, going back at least to 1895 in H. G. Wells' essay "The Limits of Individual Plasticity." He suggested, "A living thing might be taken in hand and so moulded and modified that at best it would retain scarcely anything of its inherent form and disposition; that the thread of life might be preserved unimpaired while shape and mental superstructure were so extensively recast as even to justify our regarding the result as a new variety of being" (36). The following year in his novel *The Island of Doctor Moreau*, Wells created a truly horrible world where animals were turned into pseudo-men by means of vivisection.

The genetically useless cruelty of vivisection does not stop with Doctor Moreau. It is found, peripherally at least, in "Alpha Ralpha Boulevard" by Cordwainer Smith, "Ugly red scars on his forehead showed where the horns had been dug out of his skull. He was a homunculus, obviously derived from cattle stock. I had never known that they left them that ill formed" (153). Surgery is also present in Lester del Rey's "The Faithful" (1938). Even in Clifford Simak's *City* (1952), one of the kinder visions of Uplift, a member of the Webster family says, "It was not easy, for a dog's tongue and throat are not designed to speak. But surgery did it" (60).

David Brin points out in his Afterword to *Heaven's Reach*, that works such as *The Island of Doctor Moreau*, *Planet of the Apes* (1963), and the *Instrumentality* series of Cordwainer Smith (1961 and later) "assume that human 'masters' will always do the maximally stupid/ evil thing. In other words, if we meddle with animals to raise their intelligence, it will be in order to enslave and abuse them" (431).

While it is true that many works on the subject of Uplift have been warnings against cruelty, hubris, or both, there have also been others with very different purposes. There were some that had more to do with present fun than future predictions, such as L. Sprague de Camp's stories about an uplifted bear, starting with "The Command" in 1938, or Fredric Brown's 1942 "The Star Mouse," with a mouse whose name a German scientist mispronounces as "Mitkey."

Olaf Stapledon's novel *Sirius* (1944) explores intelligence itself and the idea of something as smart as a human but different in the way its mind works, an idea that animates pretty much every examination of Uplift, fictional or otherwise, not to mention science fiction's explorations of Aliens and Robots, and nearly all the imaginary creatures in fantasy. J.R.R. Tolkien called it "one of the primal 'desires' ... the desire ... to hold communion with other living things" (15). Or in Clifford Simak's more practical SF formulation, "A *different* mind than the human mind, but one that will work with the human mind. That will see and understand things the human mind cannot, that will develop ... philosophies the human mind could not" (59).

Clearly too, Uplift is a speculation on evolution, past and future. Early stories such as Eric Frank Russell's "Mana" (1937) deal with the end of humanity and passing the torch of intelligence to successor species. In "Mana," the last man says, "Even as it was given to us by those whom we could never know, I give it to those who can never know men" (5–6). The grand gesture is undercut a little by the fact that the recipients are ants, but the great issues of Forerunners and Successors are here. So too in Lester del Rey's "The Faithful," men seek to pass on their legacy to dogs and great apes. In Simak's *City*, the inheritors are dogs and robots (since dogs lack completely effective hands in both works).

David Brin's Uplift Universe deals with beginnings, ends and the long, painful, sometimes sinister process in between. In his books, a client "is a species that owes its full intelligence to genetic uplift by its patron race" (*Startide Rising* xi). This is not, however, the sort of benevolence I have just been describing. Here, "the resulting client species serves its patron for a period of indenture to pay for this favor" (xiv).

Humans have refused to participate in such a pernicious system, but they have uplifted both chimpanzees and dolphins. This gives them status as a patron race and protects their "clients" from more predatory patrons. Additionally, humans are in the odd position of being **wolfling**s, a sentient species without a patron race so that either they achieved sentience on their own or, as most Galactic races take pleasure in believing, were abandoned by their patrons in disgust.

Brin has created one of the longer running thought experiments in science fiction, where, in somewhat the same way that Isaac Asimov examined the possible problems and advantages of building intelligent robots, Brin has been examining the problems, temptations, and possibilities of Uplift. His optimism is persuasive.

In the science fiction course I taught in the fall of 2015, almost all the students, having read *Startide Rising*, were in favor of uplifting chimpanzees and dolphins, in spite of my repeated cautions that there were already more than seven billion rather troublesome creatures on the planet who were, themselves, not much different from uplifted chimpanzees.

My students were well aware that the idea they were debating would soon be possible, in fact, in all probability, would soon be real. In the words of Enriquez and Gullens, "Almost any genetic mutation, benign or otherwise, man-made or natural, recent or ancient, from any species can be re-created and introduced into a living cell, at will, sometimes in an afternoon.... CRISPR can cut, remove, and replace hundreds, perhaps thousands of genes in the genome of a living cell per day" (134). This gene-editing technology is "in the hands of high school and college kids; it has permeated scientific research" (134). The only question seems to be how soon will we be speaking with those uplifted minds, like us but unlike, perhaps monstrous, perhaps magical but, at last, wholly other?

There are, however, a few issues left to resolve. For instance, even though we have left surgeries as a means to uplift far behind, do we have the right to mold and modify a species? In 2007, "The people of the Balearic Islands, an autonomous region of Spain, saw their parliament become the first legislature in the world to approve a resolution granting legal rights [personhood] to all great apes" (Rose). In 2013, the government of India's Central Animal Authority declared that "Cetaceans ... should be seen as 'non-human persons' and as such should have their own specific rights" (Bancroft-Hinchey).

In effect, the decisions re-defined great apes and cetaceans and preemptively removed them from various kinds of human interference, including, in all probability, Uplift. The decisions were the result of years of scientific research and political maneuvering. As far back as 1979, for instance, scientist and activist Jacques Cousteau was condemning the idea of keeping cetaceans in captivity and declaring on the subject of orcas that "on land no animal other than man himself displays such—let us say—intelligence" (168).

But do animals have the right to be left alone or is there as George Dvorsky puts it an "ethical imperative to uplift"? He argues that humans will move from "simply protecting animals" to gifting "nonhuman animals with the requisite faculties that will enable individual and group self-determination, and more broadly, to give them the cognitive and social skills that will allow them to participate in the larger social politic that includes all sentient life" (Dvorsky).

It's not hard to find people who disagree with him. Paul Raven objects to Uplift as a mechanism for making other creatures more like people. He claims, "This is the voice of assimilation, the voice of homogenisation, the voice of empire. It is the voice of colonialist arrogance, and a form of species fascism" (Raven).

Interestingly, Dvorsky's strongest argument for Uplift is a transhumanist argument, "It would be unethical, negligent and even hypocritical of humans to enhance only themselves and ignore the larger community of sapient non-human animals. The idea of humanity entering into an advanced state of biological and/or postbiological existence while the rest of nature is left behind to fend for itself is distasteful." If other species become part of such an adventure, it could well be argued that all, humans most definitely included, could transcend their previous limitations.

Finally, David Brin suggests that some animals have come up against a "glass ceiling," a barrier they may not be able to break without help. He also says, "I can tell you that the dolphins who interact with sincere human researchers appear to want—desperately—to be smarter than they are. It is a subjective impression I have heard from the scientists themselves, a number of times" ("Are Animals Intelligent"). If David Brin is correct (and this is a subject about which he has thought seriously for decades), then Uplift may prove to be a logical (if not entirely natural) next step in evolution, as the loneliness, curiosity and sympathy of the first creatures on a planet to reach sentience drive them to break the last great barrier which keeps others from joining them on that extraordinary pinnacle.

As Brin himself describes them for our own planet, "Natural beings who may not have to bump against the hard ceiling of their *pre-sapient limits* forever, but whose destinies may be broad and vast indeed ... providing we grow wise and good and skilled enough to show the way" ("Are Animals Intelligent").

Uterine Replicator (also Artificial Uterus)

A medical device designed to mimic human female reproductive organs and the functions necessary for gestation of a healthy human being. Lois McMaster Bujold coined the term in *Shards of Honor* (1986), the first novel in the Vorkosigan Saga.

Bujold gives the following description of a birth from a uterine replicator in *Ethan of Athos* (1986)—the third novel to be published in the series: "The birth was progressing normally. Ethan's long fingers carefully teased the tiny cannula from its clamp.... He checked his instrumentation: placenta tightening nicely, shrinking from the nutritive bed that had supported it for the last nine months.... Quickly he broke the seals, unclamped the lid from the top of the canister, and passed his vibrascalpel through the matted felt of microscopic exchange tubing. He parted the spongy mass, and the medtech clamped it aside and closed the stopcock that fed it with the oxy-nutrient solution. Only a few clear yellow droplets beaded and brushed off on Ethan's gloved hands. Sterility obviously uncompromised, Ethan noted with satisfaction, and his touch with the scalpel had been so delicate that the silvery amniotic sac beneath the tubing was unscored. A pink shape wriggled eagerly within.... A second cut, and he lifted the wet and vernix-covered infant from its first home" (1–2).

The idea of a device that allowed exogenesis has been around for a long time. It was mentioned as early as 1923 by J.B.S. Haldane in his Cambridge lecture "Daedalus or Science and the Future." And, of course, Aldous Huxley employs mechanisms of artificial gestation in *Brave New World* (1932). More recently, it has been used (among other places) in Frank Herbert's *Dune* (1965), Marge Piercy's *Woman on the Edge of Time* (1976), Phillip K. Dick's *The Divine Invasion* (1981), and David Weber's Honor Harrington series (1992–2016).

However, Bujold's explanations and explorations of the idea are some of the most detailed, and given the popularity of her books and the importance she ascribes to the technology (saving Miles Vorkosigan's life, for instance), some of the most important in making people aware of and comfortable with the future technology. The term she uses—uterine replicator—has been adopted by other SF writers. John Ringo, for example, uses it in *There Will Be Dragons* (2003) and in the sequel, *Emerald Sea* (2004).

In addition to introducing the technology, Bujold consistently develops its social implications, not stopping at the obvious—the liberation of the female body from childbearing—but venturing into new social structures that could be made possible with extra-uterine gestation. In the Afterword to *Miles, Mystery and Mayhem*, Bujold explains, "Primary among my beliefs was that, given humanity as I knew it, there wasn't going to be just one way any new tech would be applied—and that the results were going to be even more chaotic than the causes."

Thus, in *Cetaganda*, Bujold introduces her readers to the ruling class of the Cetagandan Empire, who use uterine "replicators and associated genetic engineering to construct their race's entire genome as a community property under strict central control" ("Afterword" 555). As a result, the people regard what we today consider natural birth as barbaric and even somewhat disgusting: "Lord Yenaro said to Lady Gelle, 'did you know that Lord Vorpatril here is a biological birth?' The girl's feather-faint brows drew in, making a tiny crease in her flawless forehead. 'All births are biological, Yenaro.' 'Ah, but no. The original sort of biology. From his mother's body.' '*Eeeuu.*' Her nose wrinkled in horror" (*Cetaganda* 30).

In *There Will Be Dragons*, John Ringo presents a society with similar attitudes, "She stood back and paced as she ran through the anatomy of the female reproductive system.... She hadn't paid much attention to the system since medical school, it was just there, as useless as the vermiform appendix that most people no longer had. With uterine replicators reproduction had all been moved *out* of the female body thank God." Another possible consequence of uterine replicator technology that Bujold entertains is "a society where women's historical monopoly on reproduction would be broken" ("Afterword" 554). This is explored in *Ethan of Athos*.

Besides introducing exogenesis as a replacement for human gestation from conception to birth, Bujold discusses (in *The Warrior's Apprentice*) the use of the technology to support a pregnancy that would otherwise become unviable. As Miles Vorkosigan—a child who would have died in his mother's womb because she was poisoned—concludes, "Lucky for me they'd imported those uterine replicators—they could never have tried some of those treatments *in vivo*, they'd have killed Mother." Miles has another interesting insight, "I wish I'd known more about this when I was a kid, I could have agitated for two birthdays, one when Mother had the caesarian, and one when they finally popped me out of the replicator" (*Young Miles* 31). This is clearly the path by which uterine replicators will be developed, as medical technology supports premature births and moves ever closer to the point of conception.

Another use for such technology is Frank J. Tipler's plan for exploring and colonizing the stars with robot probes. He suggests, "We would be able to program a von Neumann probe to synthesize a fertilized egg cell of any terrestrial species.... For humans the fertilized egg cells would have to be placed in an artificial womb—such technology is currently in the beginning stages of development—in which case the target solar system would have human beings in that system within nine months." He believes that the "children could be raised to adulthood by robot nannies" (46–48). Arthur C. Clarke has a similar system in his 1986 novel *The Songs of Distant Earth*.

In discussing the uterine replicator, Terry Johnson notes, "There are a few recent advancements bringing it a step closer to reality. An emulsified liquid blood substitute called perflubron has had some success used as a replacement for amniotic fluid for premature babies in respiratory distress." Researchers at the University of Michigan, Ann Arbor are working on developing what they call an "extracorporeal life support circuit" that will help premature infants combat problems with "organ systems including the lungs, central nervous system, and the gastrointestinal tract" (Reoma et al. 53). It is by no means a fully-functional uterine replicator, but the artificial placenta they created was able to support neonatal lambs for four hours. The technology uses "the infants' own blood pressure to drive the system" (54).

A uterine replicator that will gestate a fetus from conception to birth is further away; however, scientists at Cornell University's Centre for Reproductive Medicine are making progress. According to Robin Mckie, they "have created prototypes made out of cells extracted from women's bodies. Embryos successfully attached themselves to the walls of these laboratory wombs and began to grow."

Besides helping gestate premature infants and allowing women who cannot get pregnant and carry a child to term to become mothers, uterine replicators will make it possible to produce children who have more than two biological parents. **Multiplex parenting** has been approved in the UK, and while an embryo containing genetic material from more than two parents may be implanted in a human uterus, a uterine replicator could also be an option, depending on the gender of the parents involved. Multiplex parenting does not require parents of different genders. "Many clients for such a service would be gay and lesbian couples who could have children who are genetically related to them both" (Cook).

A fully functional uterine replicator is "still several decades away," as George Dvorsky optimistically suggests, but today's scientists already know what exactly they need to do to make this technology a reality (Dvorsky "How to Build"). A successful uterine replicator will have to mimic much more than the basic biological functions of a uterus. It will have to simulate a woman's daily activity cycle, including periods of rest, the sounds a baby hears, and the sense of touch it receives while still in the womb (Dvorsky "How to Build").

While scientists are close to approximating the necessary biological conditions, it is harder for them to reverse-engineer all the bonding mechanisms that exist between a prospective mother and her child. However, it is an inevitable progression. And as scientists succeed in supporting embryos using artificial uteruses for longer periods of time (and as they push further and further back in the development of the baby), they will eventually get to the starting point. Then, the uterine replicator will move from science fiction to reality.

Utility Fog

John Storrs Hall invented the concept in 1993. Beginning with the assumption that **nanotechnology** is already in place, it posits intelligent polymorphic material, swarm "robots the size of those 1,200-to-the-inch dots on the laser-printed page" that "can change properties so as to simulate the range of properties of ordinary matter" (Hall *Nanofuture* 189).

They're small enough to breathe in without harm and linked together in a way that makes them extremely flexible. They can turn themselves into almost anything. Fill a house with them and things get really interesting. In "On Certain Aspects of Utility Fog," Hall says, "Have an operating system that has a library of programs for simulating any object you may care to; by giving the proper command you can cause any object to appear anywhere at any time. You could carry a remote control, which might happen to be shaped like a wand." He suggests that it might be possible to move beyond wands and instrumentality to the point where the tiniest movements or other indications could trigger the Foglets, "With proper programming the Fog would almost be able to read your mind." He continues, "This combination of extreme reactivity to control and virtually limitless creative and operational ability suggest a comparison with the Krell machine in 'Forbidden Planet'" ("On Certain Aspects"). (*Forbidden Planet* is an adaptation of Shakespeare's *The Tempest* where aliens have built a system that allows them to make anything simply by thinking about it. The film has inspired scientists and futurists in various ways, just as it was inspired by them.)

Another of Hall's suggestions would take the Foglets out of the house and onto the road. "Cover the road with a thick layer of **robots**. Then your car 'calls ahead' and makes a reservation for every position in time and space it will occupy during the trip" ("Stuff Dreams Are Made Of"). Clearly, a part of all this would be what might be called the ultimate **internet of things** because once the Foglets moved out of an enclosed environment, only constant communication and interaction could make their transformations possible and safe. But that's only the beginning.

Hall says, "As long as you're covering the roads with Fog you may as well make it thick enough to hold the cars up so they can cross intersections at different levels. But now your car is no longer a specific set of robots, but a pattern in the road robots that moves along like a wave.... The appearance of the car at this point is completely arbitrary, and could even be dispensed with—all the road Fog is transparent, and you appear to fly along unsupported" ("Stuff Dreams Are Made Of").

Nor is it only cars that might become patterns on the robot background. Imagine, as Hall does, that "someone uploads his brain ... which is then downloaded into the Fog robot" (*Nanofuture* 195). This is the most extreme form of what Hall calls "Real Virtuality," and indeed, it sounds like the suggestions of intelligent software programs, where humans can exist in entirely virtual environments. (See **Cyberspace**.) The difference is that here, the real environment would be transformed into an odd mix of real and virtual by Utility Fog.

Even the early stages of the use of Utility Fog will look like magic. Of course, by then, we may be used to inanimate objects with a kind of intelligence, to an environment that protects us and obeys our commands, and to a world where we are never alone. As always with a transformational technology, there are delights and dangers to come. And as always, the creation of something so powerful will be difficult, but if and when it arrives, we will, at last, feel like magicians wielding a power that works, in whatever strange, fluid, polymorphic way, on the very real world that now frustrates and perplexes us and—so far—eludes our control.

Wolfling

The term was first used by Gordon R. Dickson in the area of culture and civilization. David Brin borrowed and changed it, and it is now more commonly employed in discussions of sentience and **uplift**. There is, of course, a clear overlap in the two meanings. In 1968, in his novel *Wolfling*, Gordon R. Dickson makes the insult the term embodies and the question it raises central issues. Humans have discovered what seems to be "a whole empire of human-occupied worlds" (5) if the physically and technologically superior aliens are indeed the same as the inhabitants of Earth.

One of those aliens defines the term wolfling: "You're a human being, all right, but one who's been lost in the woods and brought up by animals, so that you don't have any idea of what it's really like to be a human" (12). The central question that Dickson pairs with the insult is, "Maybe Earth is one of their colonies that they forgot about.... Or maybe it's just coincidence that we seem to belong to the same race" (6). By the end of the book, we discover that the humans of Earth are a **Lost Race** of the alien civilization (See also **Lost World**), and we discover that the aristocrats of that empire are entitled to genetically modify "lesser breeds of humans on their Colony Worlds" (154) at will.

For David Brin, it becomes a larger question, tied inextricably to Uplift. He gives his definition, among other places, in *Heaven's Reach*, "a derogatory Galactic term for a race that appears to have uplifted itself to spacefaring status without help, or else to have been abandoned by its patron" (447). However, Brin has not abandoned Dickson's original usage, only built on it. Toward the end of *Startide Rising*, Tom Orley says to the Gubru [large birdlike aliens] he encounters on Kithrup, "We humans are wolflings. We are feral, carnivorous, and extremely fast! Do not make me eat you" (377). Clearly,

this is an intimidation tactic, using an exaggerated version of the wolfling myth against some of the aliens who supposedly believe it.

Before the novel is over, Orley has made "Gubru jerky, Tandu strips, and flayed Episiarch" for humans and dolphins to eat (460). It's true that there's nothing else edible on the planet and that Orley killed them in self-defense, but nevertheless, he made good on his threat to eat them. I suggest a combined definition: A species or individual raised to sentience or simply to adulthood by its own efforts or with nonsentient or uncivilized help. As a result of these wild beginnings, the individual or species in question is presumed to be uncivilized or wild as well.

Ultimately, both Gordon Dickson's definition and David Brin's are about the place of humanity in the Cosmos. Did we arrive here alone, have we somehow lost our way, were we helped to achieve fully human (or sentient) status by a Forerunner, a patron or just a good Samaritan? Most of the aliens in David Brin's Uplift novels assume that it is nearly impossible for a species to achieve sentience without help. How exactly the Forerunners did it in the first place becomes for many an unexplainable, possibly divine mystery. While such a position is unscientific to say the least (and Brin certainly does not advocate it), it is not hard to understand the emotions behind such stories or to find parallels for Brin's characters. It does at least make the narrative of the rise of a species to spacefaring status go more quickly, as in Kubrick and Clarke's *2001: A Space Odyssey*.

Ironically, perhaps the best candidate for another species which helped Homo sapiens on the road to civilization is a wolfling indeed. Clifford Simak says in his nonfiction book *Prehistoric Man*, "The question ... is whether man domesticated the dog or the dog domesticated man" (cited in Wixon). As I had the dogs say in my poem "Dog Stars" in *Asimov's Science Fiction*, "We made you look up to the bowl of the sky/ By howling at the moon" (101). Stanley Coren puts it much more accurately, "Without the dog, livestock management, flock tending, and herding might never have been possible, which means that the development of agriculture as the economic base of much of human society might have been delayed or even stopped" (153–154).

That puts all of humanity in the position of Kipling's Mowgli when Akela, the old head of the Pack, tells him, "Thou art a man, Little Brother, wolfling of my watching" (170). But a man, a person, perhaps, only by virtue of that watching, of that raising to a civilization otherwise unattainable.

Part Two: Genre Terms

Alternate (or Alternative) History

A story where history takes a different turn and winds up at a very different destination, sometimes as the result of a small alteration such as Philip II surviving the assassination attempt and leading his army into Persia or Robert E. Lee accepting the command of the Union army when Lincoln offered it to him. Jeff Prucher indicates that the first use of the term was in *Fantasy and Science Fiction* in 1954 (4). Historians, non-genre writers and SF writers have all been interested in this genre (or subgenre). The first generally recognized example (1836) is Louis Geoffrey's *Napoleon and the Conquest of the World 1812–1832*. One of the earliest contributions from a historian also featured Napoleon, G.M. Trevelyan's 1907 "If Napoleon Had Won at Waterloo." That's only fair, I suppose, given how much of the world (and the future) Napoleon had rewritten.

As David Brin says, "SF writers ... devour history" (*Insistence of Vision* 18). And it's accurate to say that historians are fascinated by what might have been. It's also no surprise that scholars and SF writers find themselves thinking in similar ways about the same things. It is, not surprisingly, natural for historians to consider alternatives. After all, any criticism of one of history's blunderers (Charles I of Britain or Nicholas II of Russia spring to mind) immediately suggests that there were better choices available and that the timestream might have flowed a different way. There have been and will continue to be many collections of essays where historians examine what might have happened. In 1907, there was Joseph Edgar Chamberlin's *The Ifs of History*, which began with what is arguably its most interesting essay, "If Themistocles Had Not Beaten Aristides in an Athenian Election."

A more famous collection is John Collings Squire's *If It Had Happened Otherwise* (1931), which contains work by G. K. Chesterton and Winston Churchill. Much more recent collections include Showalter and Deutsch's 1997 *What If? Strategic Alternatives of World War II*, Niall Ferguson's 1999 *Virtual History* (with an essay from Andrew Roberts, who also edited an alternate history collection), *Third Reich Victorious* (2001), edited by Peter G. Tsouras, and Robert Cowley's 2003 *What Ifs of American History*, which lives up to the claim of its subtitle that *Eminent Historians Imagine What Might Have Been*.

On the fictional side, L. Sprague de Camp's *Lest Darkness Fall* (1939), in which a twentieth century man sets out to prevent the dark age that followed the collapse of the Roman Empire, had an extraordinary impact on SF. (See **Timeslip**.) Harry Turtledove, who is a historian and one of the best known writers of alternate history, credits the book with sparking his interest in history and its alternatives, "I was in high school when I read.... *Lest Darkness Fall*.... I started trying to find out how much he was making up and how much was real, and I got hooked" (*Agent* 5).

Indeed, curiosity is a big part of it, "Every Moment of every day, a thousand possible futures die unborn around us, a thousand corners not turned, a thousand roads not taken. We've all wondered" (Dozois and Dann). Another big part of alternate history is the desire to understand and in some small way, perhaps, to control history itself. As Brent A. Harris says in his Foreword to *Tales from Alternate Earths* (2016), "It allows us to learn more about how our world came to be by examining what could have been." Understanding is, in and of itself, a kind of control but just as much of SF tries to take possession of the future, at least as big a part wants to own the past in some way. (See **Speculative History**.) This is, of course, one of the appeals of time travel narratives of whatever type but especially of those that involve changing the past. Time travel and the quantum flux that some scientists suggest may create parallel worlds allow for almost unlimited complexities as in Philip K. Dick's *The Man in the High Castle* (plus the Amazon series based on it) and Poul Anderson's Time Patrol novels and stories.

Even non-science fiction writers have felt compelled to write alternate history. To give only a few examples: Sinclair Lewis *It Can't Happen Here* (1935), Vladimir Nabokov *Ada* (1969), Kingsley Amis *The Alteration* (1976), Len Deighton *SS-GB* (1978), and Robert Harris *Fatherland* (1992). Lewis looks at what might have happened if Huey Long had not been assassinated, Nabokov postulates a world in which (among other things) telephones are powered by water and Western Canada speaks Russian, Amis examines a timeline without a significant Protestant movement, and both Deighton and Harris suggest a German victory in World War II. (See **Realtime**.)

Archetype

The first of its kind and therefore the thing itself, a prototype or Platonic ideal. Descartes said, "And although it may be the case that one idea gives birth to another idea, that cannot continue to be so indefinitely; for in the end we must reach an idea whose cause shall be so to speak an archetype, in which the whole reality [or perfection] which is so to speak objectively [or by representation] in these ideas is contained formally."

Jungian theory uses the term in similar fashion, though the source of the prototypes is different. In Jungian theory, archetypes are patterns which emerge from the collective unconscious (residual, ancestral, mythological memories) of the human race. In *Psychology and Education*, Jung wrote, "In many dreams and in certain psychoses we frequently come across archetypal material, i.e., ideas and associations whose exact equivalents can be found in mythology. From these parallels I have drawn the conclusion that there is a layer of the unconscious which functions in exactly the same way as the archaic psyche that produced the myths" (109). In *Man and His Symbols*, Jung pointed out, "The production of archetypes by children is especially significant, because one can sometimes be quite certain that a child has had no direct access to the tradition concerned" (61).

Many people maintain that such patterns (however they originate) form an important part of myth, literature and humanness, but there are those who disagree. Roberta Rogow in *Futurespeak* defines the term more cynically, "A character or plot-line so hoary with age that it has ceased to be a stock figure or situation, and has been accepted as a myth in and of itself" (16).

Archetypes include the hero and the various parts of the hero's journey or **monomyth**, plus the wise old man, the garden paradise, death and rebirth, the innocent child, the trickster, and many others. Joseph Campbell says that for the hero's journey, "The archetypes to be discovered and assimilated are precisely those that have inspired, throughout the annals of human culture, the basic images of ritual, mythology, and vision" (18). More importantly for my purposes, archetypes often bring a sense of recognition and a powerful energy to the SF stories they help to shape.

Cassandras

In the conclusion to her essay, "Loving the Other in Science Fiction by Women," Karma Waltonen writes, "Many collections of science fiction by women reference the authors as Cassandras, as prophets whose warnings, though valid, are not heard. This comparison to the Greek heroine is perhaps more profound than we might expect.... Female science-fiction writers ... often go unheard as they write literature of social protest in an undervalued genre; they are also frequently overlooked as writers in their genre because they are female. While many of their works draw special attention to the complex matrix of sex, class, race and power, they are routinely dismissed as simply writing about gender issues if sex/gender systems are included in the story matrix at all" (42).

The essay appears in the first issue (January 2016) of the *Museum of Science Fiction's Journal of Science Fiction*, but the problems have been around for a long time. In the Introduction to her *Lost in Space: Probing Feminist Science Fiction and Beyond* (1993), Marlene S. Barr calls "feminist science fiction" a "twice marginalized field" (2). There is no question that female SF writers, futurists and scientists often get less attention and credit than their male counterparts. Nathalia Holt's 2016 book *Rise of the Rocket Girls: The Women Who Propelled Us, from Missiles to the Moon to Mars* is filled with examples.

Here is the original story of Cassandra as told by Robert Graves, "One day Cassandra fell asleep in the temple, Apollo appeared and promised to teach her the art of prophecy if she would lie with him. Cassandra, after accepting his gift, went back on the bargain; but Apollo begged her to give him one kiss, and as she did so, spat into her mouth, thus ensuring that none would ever believe what she prophesied" (263–264). Cassandra warned of the destruction of Troy and also of her own and Agamemnon's murders, but to no avail. When the ghost of Agamemnon speaks with Odysseus in the Underworld, he says about his own death and that of his entourage, "Most pitiful of all was the cry I heard from Priam's daughter Cassandra as treacherous Clytemnestra slaughtered her over me" (Homer, Shewring transl. 137).

Fortunately, the tide has turned or the exoplanets have aligned properly or humans have become at least a little more rational. Science fiction is close to becoming the literary mainstream, and the role of women in science fiction is no longer a matter for lamentations though there are continuing questions of equal treatment, equal attention and equal respect. Things are changing for the better, not only for women who write SF but for fictional women in SF. There are far more female role models than there once were (some of them created by male writers).

We've gone from Horatio Hornblower to Honor Harrington (or Kytara Vatta, Elizabeth Moon's heroine), from Kolchak to Buffy, and from Logan's run to Katniss Everdeen's target practice. There are many more, but I point out one specially interesting example from the often male-dominated worlds of space opera. *Star Trek: Voyager* was one place in Federation space where women truly had equal rights. The captain (Janeway), the chief engineer (B'Elanna Torres), the de facto science officer (Seven of Nine), the crewmember who evolves into a higher life form (Kes), the precocious child who will grow up to be a Starfleet officer (Naomi Wildman), the main villain (Borg Queen), and the most important traitor (Seska) all were female.

All the other *Trek* series combined don't equal that number of major female roles. And Janeway more than holds her own among *Trek* captains. She operates on a farther frontier than the others, and on this seven year odyssey, she becomes Odysseus, wily and violent but with her own rules.

If Picard is cautious and Kirk is dangerous, Janeway, who sometimes seems to be an authentic maniac—unbound by any archaic rules restricting female behavior—has left such conservative strategies far behind. In "Scientific Method," after plunging her ship between the suns of a binary pulsar, Janeway says to her blindly loyal Vulcan security officer, "I never realized you thought of me as reckless, Tuvok." He responds, "It was clearly an understatement."

In the two-part episode "Year of Hell," Janeway struggles with Annorax, captain of a time ship. Tom Paris, *Voyager's* irreverent helmsman, labels him "Captain Nemo" and says of his behavior, "That's called paranoia ... with a hint of megalomania." But Janeway is equally unyielding. She states, "We're going through their space whether they like it or not." She fights her ship for the year of the title until it is not much more than wreckage, and then, alone on *Voyager*, she destroys Annorax, kills herself, and restores the timeline by ramming the time ship.

In the long arc of *Star Trek's* storylines, the partly mechanical, partly biological Borg are the Federation's most terrifying enemies. Only Janeway regards their powerful ships as good places from which to take technology by force. She makes deals with the Borg, steals Seven of Nine from them, and comes back from the future as Admiral

Janeway to crush the Borg Queen one last time. In *The Farther Shore*, a *Voyager* novel set after the end of the series, Christie Golden has Janeway say, "The Borg are so familiar to us, they're like old friends" (31). In Janeway's universe, they are more like old targets, easy to hit after long practice.

But *Voyager* is not about the power and freedom of one person, however indomitable. *Voyager*'s journey is, as all good quests are, a quest for humanness. Seven of Nine, a human assimilated by the Borg when she was six and forcibly rescued by *Voyager* eighteen years later, reluctantly abandons her Borg nature and struggles to rediscover and reinvent her humanity. Or as the title of *Star Trek Scriptbooks Book Two* puts it, *Becoming Human: The Seven of Nine Saga*. In "The Gift" Janeway says of the appeal of the Borg, "You were part of a vast consciousness, billions of minds working together, a harmony of purpose and thought, no indecision, no doubts, the security and strength of a unified will." In this description, the Borg seem to be more than unity; they are approaching divinity.

Seven complains, "This drone is small now, alone, one voice, one mind, the silence is unacceptable." But she fights her way from nonentity to identity, from despair to hope. In the process of finding herself, she questions the assumptions and values of her crewmates. Though Janeway is her mentor, Seven finds the strength to question and even defy her, no mean feat where Janeway is concerned. In "Prey," Seven, who has just disobeyed the Captain's orders and saved the ship, complains, "You made me into an individual.... You encouraged ... my independence and my humanity, but when I try to assert that independence, I am punished."

Janeway responds, "Individuality has its limits. Especially on a starship where there's a command structure." But Seven is not easily silenced, "I believe that you are punishing me because I do not think the way that you do, because I am not becoming more like you." It is a continuing argument, an engagement with the self who is also the other, that leads to enlightenment for them both, just as Seven's journey to humanness and hope illuminates the transformations of the other crewmembers and their joint struggle to get to that place of hope which is home. (An earlier version of this appeared in Pilkington "*Star Trek*" 63–65.)

In the end, the term "Cassandras" has more resonance than might at first appear. Cassandra was a legendary figure who became a literary one. While in the legend, her prophecies were ignored, in Homer's version they were highlighted. Agamemnon, who took her as a war captive, finds the end of her story more compelling, more worthy of pity than his own death, though he has plenty of pity for himself as well. Maybe Samuel Butler and Robert Graves were right when they argued that the author of *The Odyssey* was a woman.

It is good to remember female science fiction writers, scientists and futurists, and their prophecies, and good to remember that some of them were heard. Some of them wrote authentic self-defeating prophecies, warnings that worked, from Rachel Carson to Jane Goodall. Of course, the list of truly extraordinary female writers of SF is too long for one of my entries. Here's a remarkable but vastly incomplete one: Eleanor Arnason, Leigh Brackett, Marion Zimmer Bradley, Octavia Butler, Lois McMaster Bujold, C. J. Cherryh, Nancy Kress, Ursula K. Le Guin, Anne McCaffrey, Vonda McIntyre, C. L. Moore, Andre Norton, Joanna Russ, Mary Shelley, Sherri S. Tepper, James Tiptree, Jr. (Alice B. Sheldon), and Chelsea Quinn Yarbro. Their works will still be read long after the prejudices I've been discussing have become small footnotes in large books. Moyle Rice, one of my undergraduate professors at Utah State University, said, "Never make war on a people with great writers. They'll defeat you forever."

Fantasy

The *OED* provides a first example in print in 1949. That strikes me as a bit late since as David Brin says, "Up till the early 18th century ... nearly all previous storytelling contained elements of the fantastic" (*Insistence of Vision* 16). But ideas often predate words, and we would have to discuss how much of what we now think of as fantastic in *The Odyssey*, say, was considered fantastic or untrue when Homer was first chanting it. In short, as with **science fiction**, there is more argument than agreement about the term fantasy. Clute and Grant in their *Encyclopedia of Fantasy* use a few more words to get to the same place, "Fantasy's specific location in the spectrum of the fantastic is a matter of constant critical speculation" (337).

Perhaps the simplest definition is a distinction between the two related and often overlapping genres—science fiction is about what doesn't exist but could, while fantasy is about what never has existed and, by the laws of our universe, never will. This is, of course, to define all of science fiction as ***hard sf***.

At its largest, a definition of fantasy could engulf all fiction, because, after all, fiction is inherently unreal. A slightly smaller circle would put the fantasy label on all non-mimetic fiction, that is, all fiction which does not strive for verisimilitude or try to recreate consensus reality (the world as we believe it truly is, and yes, I know there are many disagreements about the nature of reality, but essentially I'm talking about the scientific world view).

This definition, which can be reduced to fantasy is fiction about the world as it isn't, inevitably blurs the distinction between Tolkien and Asimov. For instance, *The Concise Oxford Dictionary of Literary Terms* (1990) defines fantasy as, "a general term for any kind of fictional work that is not primarily devoted to realistic representation of the known world. The category includes several literary genres (e.g., dream vision, fable, fairy tale, romance, science fiction) describing imagined worlds in which magical powers and other impossibilities are accepted" (Baldick 81–82). Or in the words of *The Encyclopedia of Science Fiction*, "All sf is fantasy but not all fantasy is sf" (Clute and Nicholls 408). To add to the confusion, various postmodernist (such as Kurt Vonnegut and Thomas Pynchon) and magic realist (e.g., Gabriel García Márquez, Günter Grass, and Salman Rushdie) writers have employed the devices of fantasy to further their own literary ends.

Ultimately, the most useful distinction between fantasy and SF may be a philosophical one: SF envisions a world (and a self) that can eventually be known and controlled, a universe within the grasp of the human understanding. Fantasy, on the other hand, sees a mystical, inexplicable, sometimes glorious, sometimes terrifying universe both within and without. Science fiction is far more likely to present characters whose choices create their lives, while fantasy characters walk preordained paths; their destinations are also destinies. To put it another way, science fiction is about knowledge; fantasy is about power.

If all these definitions seem a little vague, here are some additional specifics: SF is about science, while fantasy deals with magic. SF happens much of the time in the future, while fantasy is more likely to occur in the past. A depiction of a society with equal rights and opportunities is more likely to be science fiction, while kings, knights and fixed hierarchies tend to be fantasy. And finally, if you can comfort yourself by saying that the story you're reading or watching could never happen, it's fantasy. If you're either worried that what you're reading about will not come true in your lifetime or afraid that it might, it's SF. (An earlier version of this appeared in Pilkington "Introduction" *The Fantastic Made Visible* 7–10.)

Faust

The name has become both a symbol and a battleground. Faust was a scholar who gave up his soul to gain knowledge. There are many candidates for the original of the character (including Simon Magus), but for what it's worth, Christopher Marlowe's source was an English translation of the "German *Historia von D. Johan Fausten* or 'Faustbuch' first published ... in 1589" (Ribner vii). As the story has changed over the centuries, Faust has become the quintessential (and often oversimplified) example of the human who gives up his soul to gain power, knowledge and pleasure or some other combination of earthly temptations.

The various Faust stories are ultimately based on a real 16th century German philosopher who claimed to have magical powers. As Brian Stableford puts it in *Science Fact and Science Fiction*, "After the actual Faust's death, the rumor was spread that he had traded his soul to the Devil in exchange for 'earthly knowledge'; his career thus became a parable in which science is represented as essentially satanic by virtue of its concentration on the empirical at the expense of the spiritual" (177). (See **Frankenstein Complex**.)

The story has changed over the centuries and become a sort of litmus (or perhaps Rorschach) test, with multiple interpreters finding various meanings so that Faust sells his soul for a number of different things. In William Tenn's "Bernie the Faust" (1963), Bernie says, "I was a new kind of Faust, a twentieth-century American one. The other Fausts, they wanted to know everything. I wanted to own everything" (17). Some of the more famous versions of the story are Christopher Marlowe's *The Tragical History of Doctor Faustus* (around 1592), Goethe's *Faust* (Part I 1808, Part II 1832), Byron's *Manfred* (1817), Thomas Mann's *Doctor Faustus: The Life of the German Composer Adrian Leverkühn as Told by a Friend* (1947), and Bulgakov's *The Master and Margarita* (completed sometime around 1940 but not published until 1967). Christopher Marlowe's version in many ways set the pattern for much of what came later. While there is plenty of disagreement on the subject, "*Faustus* has been viewed as an agnostic protest against the limitations imposed by Christianity upon the normal aspirations of the human spirit, with Marlowe the apostle of an aspiring Renaissance humanism" (Ribner vi). However, it is extremely difficult to evaluate (or even to understand) a sixteenth-century tale about a deal with a devil without some sense of what Marlowe and his audience believed or at least accepted as background assumptions when they went to a play.

The historical context of Marlowe's *Doctor Faustus* and similar works written at the same time makes clear how complicated the issues are and how far removed from a simple trading of one commodity for another. The historical context also paints a picture of the playgoers as a sophisticated, skeptical bunch who were inclined to accept some magic (and magicians) as perfectly legitimate—at least on the stage.

Renaissance theorists (and playgoers) recognized at least four different types of magic. The first and the most innocent was not essentially different from protoscience. Though its basic assumptions about the magical affinities within nature proved to be wrong, it was still an attempt to manipulate nature in natural ways: "No invocations are offered, no implorings made; whatever consciousness exists in non-human nature is not constrained by ceremonies to be helpful" (Shumaker 108). So, Friar Laurence, in Shakespeare's *Romeo and Juliet*, gathers herbs at dawn, saying, great "is the powerful grace that lies/ In plants, herbs, stones, and their true qualities" (2.2.15–16).

This vision of the world has obvious similarities with alchemy, and if Subtle (in Ben Jonson's *The Alchemist*) had refrained from meddling with "familiars," he might have been able to claim as Giovanni della Porta actually did in his *Magiae naturalis libri viginti*, that "none of the information offered ... is in any way illicit" (Shumaker 110). It's worth remembering how much time Isaac Newton, the greatest scientist of his age, spent on alchemical activities—enough to make him an expert on metals and coinage when he became Master of the Mint.

The archetypal white magician, or mage, though his intentions might have been equally innocent and his life was supposed to be one of exemplary purity, stood on more dangerous ground and defended more debatable doctrines. The complex issues of Renaissance pneumatology swirled round his head like a troubled nimbus. Since he used ceremonial magic and called up spirits, he was plunged into the center of the controversy; it is a labyrinth with no single thread to follow to safety.

Those writers who believed with Cornelius Agrippa that "good daemons can be attracted and bad ones repelled" (Shumaker 151–2) were willing to accept the white magician on his own terms, as a Neoplatonist philosopher who "sought to refine his soul and gain a direct knowledge of God" (Woodman 30). In this view, a creature like Ariel in Shakespeare's *The Tempest* is not an evil demon sent to damn human souls but, as C. S. Lewis puts it in *The Discarded Image*, a member of "a third rational species distinct from angels and men" (134), which served as a bridge between them.

So, in the *Star Trek: The Next Generation* episode "Emergence," Data, who is playing Prospero, *The Tempest*'s great wizard, on the holodeck (and who represents in himself another rational species), responds to Captain Picard's criticism by saying, "I am supposed to be attempting a Neoplatonic magical rite."

Whatever the confusions of the controversy or the level of theological sophistication in theatre audiences, it is clear that there was a place for the white magician in English Renaissance drama. In a discussion of positive reactions to Doctor Faustus from other (necessarily mistaken) characters in that play, Anthony Harris argues that, "Such an attitude is in accord with the spirit of the romantic comedies of the early sixteenth century, where wizards and enchanters were honoured and the legality of their magical practices was unquestioned" (117). David Woodman maintains that "most audiences possessed such a truly commonplace knowledge of magic, both black and white, that a popular response to Prospero as a white magician was assured" (73).

The black magician is easier to classify because there was no doubt about the nature of the spirits he summoned. In their discussion of the relative merits of pyromancy and geomancy, Friar Bungay and Jacques Vandermast (from Robert Greene's *Friar Bacon and Friar Bungay*) take it for granted that the spirits with which they deal are demons. Vandermast says, "When proud Lucifer fell from the heavens,/ The spirits and the angels that did sin with him/ Retain'd their local essence as their faults" (1.9.58–60). He goes on to list their local habitations, filling fire, air and earth with fiends.

Of course, any kind of contact with demons was dangerous; their influence was supposed to be malign, their intentions deadly, and their essential natures damnable. The magician might choose to use such diabolical means to achieve his virtuous ends (as Friar Bacon does), but he hazarded his life and his soul each time. The knowledge that he acquired was forbidden; the power that he took was unchancy and treacherous. And there was always the risk that he would follow Doctor Faustus' way to the everlasting bonfire by becoming a witch.

The witch and the black magician were, however, very different kinds of creatures, at least as the mythology came to be laid out and as the deal with the devil "had been adapted as an important instrument of the politics of persecution" (Stableford *Science Fact and Science Fiction* 177). The magician maintained his independence, calling demons by means of his knowledge and compelling them with the force of his will. He stood on the edge of blasphemy, but he could still freely draw back. The witch, on the other hand, was already damned, having gained "occult powers through a formal pact with the devil" (Harris 3).

Thus, the witch worked magic, not through wisdom and skill but as a result of an agreement with Lucifer or one of his representatives, and the inevitable end to this agreement (and the only reason for such services as the demons performed) was the surrender of the victim's soul.

The witch might be an old, bedeviled woman who turns finally, as Mother Sawyer does (in Dekker, Ford, and Rowley's *The Witch of Edmonton*), to the devil himself as the only possible source of comfort, or less often, the witch might be a scholar like Faustus, whose dreams and impatience plunge him into hell. In Marlowe's play, Faustus asks Mephistophilis, "Did not my conjuring raise thee? Speak." And Mephistophilis indicates that Faustus is on the way to witchcraft and damnation, "That was the cause, but yet per accidens,/ For, when we hear one rack the name of God,/ We fly in hope to get his glorious soul" (1.3.46,47,49). [This is very much like the dog-demon's line in *The Witch of Edmonton,* "Ho! Have I found thee cursing? Now thou art/ Mine own" (Dekker et al. 2.1.118).] It is not the power of Faustus' spell but the hope of his damnation that draws Mephistophilis. As he says, "I came now hither of my own accord" (1.3.45).

Witch, wizard, black magician: all are symbols of magical power. Since such power is dangerous at its best and damnable at its worst, those characters who use it become heroes extending human boundaries, rebels challenging the established order, and aliens standing solitary and apart from their own communities. Much of the energy behind these themes comes from the magical view of the universe, which places man at the center, the microcosm mirroring the macrocosm, personally affected by the animate forces of stars, spheres and spirits, and able by his actions to affect them.

As Jeffrey Burton Russell puts it in *Witchcraft in the Middle Ages,* "Magic is a doctrine that, far more than religion or science, exalts man to the loftiest regions of glory: hence its perennial attraction, and hence its particular appeal for the Renaissance, when man's ambitions and his ability to achieve them seemed unlimited" (5). Magic in English Renaissance drama becomes a metaphor for the immensity of man's dreams, the heights to which he may possibly climb and the depths to which he could conceivably fall.

The protagonist who extends human boundaries by means of magic is part of a long mythological tradition, and is also part of a new recognition of the importance of human knowledge and discovery. As Joseph Campbell describes the process in *The Hero with a Thousand Faces,* "The hero goes forward in his adventure until he comes to the 'threshold guardian' at the entrance to the zone of magnified power. Such custodians bound the world.... Beyond them is darkness, the unknown, and danger" (77). The demons and daemons that witches, wizards, and black magicians confront are such "threshold guardians," marking the limits of human society.

Doctor Faustus' blithe disregard of damnation, "I confound hell in Elysium./ My ghost be with the old philosophers" (1.3.60–1); Mother Sawyer's reluctant acceptance of the pact with the devil, "When I am thine, at least so much of me as I can call mine own" (2.1.141–2); Friar Bacon's resolve to "turn my magic books/ And strain out nigromancy to the deep" (1.2.53–4); Prospero's "being transported/ And rapt in secret studies" (1.2.76–7); and even Sir Epicure Mammon's, "This is the day wherein, to all my friends,/ I will pronounce the happy word, 'Be rich'" (*The Alchemist* 2.1.6–7)—are all ventures across a threshold into another world.

Both Friar Bacon and Doctor Faustus are presented as culture heroes who expand human horizons, performing amazing and admirable feats. Faustus is, of course, a sinner caught in the same web as Mother Sawyer, but as David Adams Leeming puts it, "more important is the human yearning to break free which is behind Faust's sins" (186).

Both Bacon and Faustus are part of the tradition of the magician who is cloaked in good intentions like Owen Glendower in *Henry IV, Part 1*, Merlin in *The Birth of Merlin*, the resurgent Bacon himself in *John of Bordeaux* and, most obviously of all, Prospero in *The Tempest*. If the hero can benefit society greatly enough, there may be a sanction for all his actions. Thus, Leontes in *The Winter's Tale* declares, "If this be magic, let it be an art/ Lawful as eating" (5.3.110-111).

And it is in this context of a society in transition, of culture heroes who bring with them the new culture of the Renaissance or Enlightenment or Romanticism (or even a strange amalgamation of all three) that Faust emerges as a hero. He is glorious even when he fails because he shows the power of humanness. The audience that came to see Marlowe and Shakespeare knew which side they were on, as did the audience that read Goethe and Byron.

Lord Byron makes the issue abundantly clear in *Manfred*. As with any good Faust story, the spirits arrive at the end to summon Manfred to hell, but his response is very much his own: "I do not combat against Death, but thee/ And thy surrounding angels; my past power/ Was purchased by no compact with thy crew,/ But by superior science—penance, daring,/ And length of watching, strength of mind, and skill" (3.4 112-116). So, Manfred, who from the beginning has called up spirits by "a tyrant spell.... The thought which is within me and around me" and not by any pact with demons, successfully defies that terrible summons (1.1 43-48). When the spirits mention his "many crimes," he responds, "What are they to such as thee?/ Must crimes be punished but by other crimes,/ And greater criminals?—Back to thy hell!/ Thou hast no power upon me, *that* I feel" (3.4 122-125).

And indeed it is his feelings, his powers—himself—this Manfred as the measure of all things, who denies any right of the supernatural to control him either alive or dead. He brushes off the Abbot as readily as he sweeps aside the evil spirits. In true Byronic style, he declares, "What I have done is done; I bear within/ A torture which could nothing gain from thine:/ The Mind which is immortal makes itself/ Requital for its good or evil thoughts,—/ Is its own origin of ill and end—/ And its own place and time" (3.4 127-132).

It may well be that Byron has said at least a few of the things that Marlowe wished to say and would have said had he lived in a later time. Manfred's last words to the demons are: "*Thou* didst not tempt me, and thou couldst not tempt me;/ I have not been thy dupe, nor am thy prey—/ But was my own destroyer, and will be/ My own hereafter—Back, ye baffled fiends!/ The hand of Death is on me—but not yours!" (3.4.137-141). Manfred's last words to anyone are to the Abbot, "Old man! tis not so difficult to die" (3.4.151). The line was controversial enough that the publisher omitted it in the first edition, but Byron, being Byron, had it put back later.

Brian Stableford argues that "Like the Romantics before them, genre science fiction writers mostly sided with

Faust against Mephistopheles, convinced that the quest for knowledge was a sacred one no matter how much fonder a jealous God might be of blind faith" (*Science Fact and Science Fiction* 178).

But not everyone who has chosen to rewrite the story of Faust has wanted to tell a tale of a hero who challenged the culture of authority and belief and championed a culture of skepticism and science. And not everyone who has used the word "Faust" has intended it to mean something complex and controversial. It has been oversimplified into the condemnation of an obviously bad choice. It has, for instance been repeatedly deployed against physicists working on nuclear weapons and nuclear power, with the underlying assumption that any deal that results in great power must by its nature be wrong. (See **Atomics**.)

Fictions of Nuclear Disaster

Are novels, films or similar works about the destruction caused by a nuclear device, usually a bomb or power plant but sometimes, in the words of the WOPR computer in *WarGames*, "global thermonuclear war." Fictions of nuclear disaster are most often meant to be self-defeating prophecies, though it's hard to measure their impact with any precision.

One of the better examples comes from then President Ronald Reagan. He writes in his autobiography, "This was part of the entry in my diary October 10, 1983: I ran the tape of the movie ABC is running Nov. 20. It's called 'The Day After' in which Lawrence, Kansas, is wiped out in a nuclear war with Russia.... My own reaction: we have to do all we can to have a deterrent and to see there is never a nuclear war" (585). He goes on to say that "not long after that" he had a "briefing on our complete plan in the event of a nuclear attack.... In several ways, the sequence of events described in the briefings paralleled those in the ABC movie. Yet there were still some people at the Pentagon who claimed a nuclear war was 'winnable.' I thought they were crazy" (585–586).

I am not, of course, arguing that Reagan based his nuclear policy on *The Day After*. He had suggested eliminating all nuclear weapons as early as 1982 (Anderson and Anderson 94). It's clear, though, that he was (like many other people) influenced by the film.

The size of the disasters ranged (in the 1950s when the subgenre first became popular) from local to global or, in at least one case, even larger. Fictions of nuclear disaster were everywhere at the time. In 1942, three years before the first atom bomb and twelve years before the world's first "nuclear powered electricity generator began operation" in Obninsk, Russia ("Outline History"), Lester del Rey wrote a novella (published in *Astounding*) about a meltdown at a nuclear power plant.

Hope and fear grew together at extraordinary speed, and movies grew with them. Of the "more than 500 science-fiction features" made in Hollywood "between 1948 and 1962" (Waldman), a surprising number dealt with atomic energy directly, indirectly, or symbolically. In 1951, in *The Day the Earth Stood Still*, aliens gave humans the choice between abandoning nuclear weapons or being destroyed by a sort of police force of angry **robots**. In 1953 in the suspense film *Split Second*, a murderer who learned to kill as a soldier in World War II, takes

Fictions of Nuclear Disaster   167

hostages within the blast radius of an imminent atomic test. The explosion is part of the movie, with "The End" written across the mushroom cloud.

In the years immediately before the release of *Forbidden Planet*, nuclear explosions supposedly created an entire menagerie of mutated monsters in low-budget films, including *Beast from 20,000 Fathoms* (1953), *Godzilla* (1954), *Them!* (1954), *Creature from the Black Lagoon* (1954), *Tarantula* (1955), and *It Came from Beneath the Sea* (1955), to mention only the most notable examples.

When John Adams wrote the music for the 2005 opera *Doctor Atomic* (a very compressed version of the story of Los Alamos), he paid those 1950s films a major compliment, "When I was a little boy, I remember seeing science fiction movies. They usually had some terrible thing that'd gone wrong, some kind of pollution, and it was always the result of a nuclear bomb ... a test in Nevada or something ... and they would produce a monster or a sickness or people would become very strange. And so I thought that the science fiction aspect was something I wanted to bring in, that 1950s sense of doom and gloom, but I didn't want to do it in an ironic way. I actually wanted to treat that kind of ... film music that you see in science fiction music as a kind of **archetype**."

But fictions of nuclear disaster, in spite of their raw power, are difficult to turn into workable films. Set in postapocalyptic worlds, the stories become trivial sex farces (such as Pat Frank's 1946 novel *Mr. Adam*, where a massive explosion of nuclear power plants renders all male humans except one impotent), implausible character studies, or scenarios of destruction with survivors so alien that there is no understanding their actions or sympathizing with their motivations, like the mutated humans in *Beneath the Planet of the Apes*, who stand before a hydrogen bomb with a cobalt casing and chant, "Glory be to the bomb and to the holy fallout." (See **Doomsday Machine**.)

Set in worlds on the verge of apocalypse, the stories become didactic, political and top-heavy, weighed down with presidents, senators, generals and KGB officials. Even the James Bond films (and books), whose narratives have teetered repeatedly on the verge of thermonuclear war, suffer from some of these problems. From his excellent book *Fictions of Nuclear Disaster*, here is David Dowling's expression of the difficulties and possibilities, "We live in an age of constant threat, of potential apocalypse. The magnitude is beyond our reckoning, the technology and perhaps the politics beyond our ken, but what can be explored and dramatized is what it is like to feel in the post–1945 world.... [T]he end towards which we drive insanely is not known, only known about" (11).

Postapocalyptic scenarios were not especially common in 1950s sci-fi, though there is, for example, *World without End*. It dealt with the long-term aftereffects of an atomic war. In most films, irradiated monsters stood in for the complications of nuclear confrontations between nations, but the resulting plots could still be unwieldy. Here's Susan Sontag from "The Imagination of Disaster" describing what she sees as a "typical science fiction film" and identifying problems that are similar to the ones I mentioned: "In the capital of the country, conferences between scientists and the military take place.... A national emergency is declared. Reports of further atrocities. Authorities from other countries arrive in black limousines.... This stage often includes a rapid montage of news reports in various languages, a meeting at the UN, and more conferences" (210).

In spite of the difficulties, there were many successful, even extraordinary films. I include detailed comments here on three, not because there are no others that are comparable but because each of them found a remarkable approach to the subject of nuclear disaster, a way of illuminating the problem that might, indeed, have made them self-defeating prophecies.

Forbidden Planet has a kind of atomic monster, escalating casualties, and what might be called (without stretching the term too far) the aftermath of a nuclear conflict, which in this instance, exterminated the population of an entire planet. And for good measure, there's a thermonuclear explosion that destroys everything within 100 million miles—arguably the largest blast radius up to that point in movie history. All of this comes without government officials or governments, without crowds or newscasts. Even the military officers have no superiors, at least none that they can reach during the course of the film. *Forbidden Planet* is set entirely in space or on an alien planet, and that is part of what makes it such an effective fiction of nuclear disaster because it limits and concentrates the storyline and the characters.

The flying saucer that Adams and his crew arrive in is nuclear powered. When they set out to improvise a transmitter that they hope will allow them to communicate with Earth "they have to unship the main drive to juice it." And for this, they need "two-inch lead shielding." The makers of *Forbidden Planet* seem to have had a better idea of the dangers of radiation than some other filmmakers did at the time. "The script also called for helmets and radiation armor to be worn by the crew as they unshipped the saucer's main core ... but this special costume idea was scrapped when the film began to go over budget" (Clarke and Rubin 28).

Robby even asks Alta, when she requests he make a dress for her "where absolutely nothing must show," if it should also be "radiation proof." Of course, the nearly limitless Krell power is also atomic in nature, a reminder that nuclear weapons were not the only danger from **atomics**. Morbius describes it as, "9200 thermonuclear reactors in tandem. The harnessed power of an *exploding* planetary system" (emphasis mine).

It is, therefore, nuclear power that makes the Monster from the Id possible. As Commander Adams says to Morbius, "Look at your gauges. Look! That machine is going to supply your monster with whatever amount of power it requires to reach us." In effect, the Monster from the Id is a symbol of nuclear war, like the mutations in other films but also unlike them. The monsters in other films are examples of strange and unusual things, of the Other, of creatures transformed by radiation. It is possible for the audience to distance themselves from the monsters. Indeed, it is difficult to do anything else. But the Monster from the Id is different. He is, like Shakespeare's Caliban, the dark side of humanness or the **alter ego**, he is not meant to be the result of radiation but the cause of nuclear war. The shining

Krell, with their nearly godlike power, also carried dark passengers, and ironically and tragically, a race whose knowledge was all but limitless died of ignorance.

As I said in "Fighting the History Wars on the Big Screen," *The Terminator* is Cameron's extraordinary metaphor for the reality of nuclear war. After the explosion of the first atomic bomb, Robert Oppenheimer, who spoke six languages, including Sanskrit, quoted the *Bhagavad Gita*, "I am become Death, the shatterer of worlds" (DeGroot 64–65). Cameron's version of Death may be smaller, but it is also deadly and far harder to ignore than the massive missiles slumbering in their silos. And its ultimate outcome, Cameron suggests, may be the same—the destruction of humanity.

Oppenheimer himself was to be disgraced, in large part, because of his opposition to Truman's top secret development of the hydrogen bomb, which Oppenheimer thought "would be a weapon not of warfare but, quite possibly, of genocide" (McMillan 3). In Cameron's story, the representative of thermonuclear war goes door to door, killing a few in the present so that all may die in the future. But the terror that is attached to the Terminator is always, for the audience at least, partly the terror of thermonuclear war. The audience, unlike Sarah Connor, knows from the film's first establishing shot and voice-over that there was (or will be) a nuclear war in the future, and that the Terminator is a representative of the intelligent machine that caused it.

The Abyss is another of Cameron's extraordinary fictions of nuclear disaster, one that references the Cuban Missile Crisis, sets its events some eighty miles from that island, and uses as its background a similar standoff between the superpowers. This is the larger historical context, and throughout the film, we see what is happening, as the tension ramps up, on television screens. Even the aliens in the abyssal trench are watching. Though historical ships and shipwrecks are never far from Cameron's mind, he also uses (at least as a starting point) sonar echoes from movies such as *The Bedford Incident*, *The Enemy Below* and *Run Silent, Run Deep*, with perhaps a ping for Jules Verne.

While the sinking of a U.S. nuclear submarine is the beginning of the story, and the question of how it sank (and if the Soviets sank it) remains at the center of the narrative, *The Abyss* rapidly morphs into something entirely different. Lindsey's sarcastic questions mark the point where all doubts of an alien presence must be abandoned, "So, raise your hand if you think that was a Russian water tentacle. Lieutenant? No, well. A breakthrough."

Interestingly, Cameron has not really changed subgenres; he has just decided to construct two fictions of nuclear disaster instead of one. The first is a nuclear submarine at the bottom of the ocean which precipitates a nuclear crisis, complete with difficulties between the crew of the hastily impressed "salvage" ship and the military contingent they are forced to take with them. The second is an advanced alien civilization that decides to take action against the dangerous follies of humanity before things get "out of hand" and their own environment, inhabited for longer than humanity has been in existence, is harmed. They seem to agree with John Connor's pet Terminator from *Terminator 2*, "It is in your nature to destroy yourselves." And, in what is very likely a nod to *The Day the Earth Stood Still* (and "Farewell to the Master," the more sophisticated story behind it), the aliens provide humanity with a choice, "live in harmony or face obliteration" (Keegan 86).

As context for this choice, fictions of nuclear disaster usually contain at least some condemnation of human beings. Mutated monster scenarios may move that condemnation back one step since the immediate destruction can be blamed on the monster, and in the same way, the *Terminator* films move the blame back two stages, to the computer which sent the killer robot and the killer robot itself. But the blame is almost always there, and humanity's responsibility is seldom forgotten.

Planet of the Apes, for example, might have left the condemnation in the murky history of an Earth dominated by apes. Instead, it is made graphically clear. When Taylor (Charlton Heston) discovers what's left of the Statue of Liberty, he says, "We finally really did it. You maniacs! You blew it up. Ah, damn you, God damn you all to hell!" Ultimately, that is the choice which all fictions of nuclear disaster from *The Martian Chronicles* to *WarGames*, from *Dr. Strangelove* to *Deterrence* offer: find a means of living together or face the fatal consequences of a technology that was designed to destroy worlds.

Here is Ray Bradbury's version of that destruction in a Martian story from 1950 that wasn't included in *The Martian Chronicles*, "Only six hours ago, Jones, Williams and himself were provisioning their rocket to return to Earth. They were the last ones on Mars. They had wondered if the atom war was as bad as the space radio claimed. They had joked about it. They had figured to be in New York in a month. And then— that violent blue flash in the sky. Earth, burning, a new small sun, had set over the horizon" (106).

Fixup

Invented by A.E. Van Vogt to describe his 1975 work *Reflections*, the word means a novel made from previously published stories, almost always with some transitional or new material added. As usual, the practice predates the term, and the most famous fixup in SF history is Isaac Asimov's Foundation sequence, whose first three novels (all fixups) won the Hugo for "Best All-Time Series." The first volume (*Foundation* 1951) also included Asimov's most famous epigram, "Violence is the last refuge of the incompetent" (*The Foundation Trilogy* 66). The term is an indication that while art may imitate life, life's complications often dictate the forms art takes.

Hard SF

Science fiction that extrapolates from science and provides rational (and often detailed) technical explanations for its wonders. In James Gunn's definition, "Science fiction based on accepted scientific principles and reasonable extrapolation of those principles into the future." He adds, however, "The way science is expanding its horizons of the possible, hard SF seems limited only by the imaginations of its creators" ("Introduction" *Tomorrow and Tomorrow* iii).

Nancy Kress says, "Hard SF has several varieties, starting with really hard, which does not deviate in any way from known scientific principles in inventing the future.... However, even the hardest

SF involves some speculation or else it would not be science *fiction*" (cited in Wilde).

According to Brian Stableford, the term originated as part of an extended argument about what was (and had been) happening to *science fiction*. The complaint, aimed at "John W. Campbell Jr.'s editorial policies in *Astounding Science Fiction*," was that what J. R. Pierce called "the hard scientific and technological core" had been steadily eroded. P. Schuyler Miller quoted Pierce's article and came up with two terms "hardshell" in 1952 and in 1957 (in the February *Astounding*) "hard science fiction" (*Science Fact* 226–227). Stableford makes some large claims for Hard SF, "Hard science fiction aspires to serve as a medium of thought experiments, significant not merely as a means of popularizing scientific ideas but also—and primarily—as a means of their philosophical investigation" (228). As this book makes clear, I tend to agree with those claims, but I'm also convinced that less rigorous forms of the genre (and even other related genres) can sometimes claim to do the same things.

George Zebrowski provides an even more rigorous definition in his "Afterword: Reality Check" to *The Killing Star* (1995), a novel he co-wrote with Charles Pellegrino: "What is certain is that we must think about what we are doing, in the light of what is possible and probable. And that is the bottom line, grim reality that distinguishes 'hard' science fiction from every other kind of writing: not that something is merely possible and interesting, or metaphorical, or even aesthetically pleasing, but that it is possible, probable, even likely; that it does not exist only in the imagination, but might confront us in our daily lives—as, for example, an asteroid strike that might destroy most of civilization" (340).

As is usual with genre definitions, the arguing is at least half the fun. On one hand, the lines blur, and it's easy to include most people, but on the other, it sometimes seems as though no one is pure enough, and besides, writers are not consistent. The early H. G. Wells could not possibly have qualified, but later on, he made successful predictions, always a badge of honor in Hard SF. Here is a very short (and entirely incomplete) list of writers who are generally recognized as belonging in the pure column (at least for some books): Poul Anderson, Catherine Asaro, Isaac Asimov, Greg Bear, Gregory Benford, James Blish, David Brin, C. J. Cherryh, Arthur C. Clarke, Hal Clement, Michael Crichton, James Gunn, Nancy Kress, Cixin Liu, Linda Nagata, Larry Niven, Charles Sheffield, Jules Verne and Timothy Zahn.

The two most talked about examples of Hard SF as I was working on this book were Andy Weir's *The Martian* (2011 for the novel, 2015 for the film) and Kim Stanley Robinson's *Aurora* (2015). For those who want to be extra strict in their definitions and exclusions, I point out that Asimov's creation of the **positronic brain** was much closer to *fantasy* than to physics, and yet it may turn out to be, in its own nonscientific way, one of Asimov's most successful prognostications, not because he provided a blueprint but because he set a nearly irresistible goal.

Lost Colony

In SF, a colony (usually on a distant planet) that has lost touch with Earth (through chance, neglect, collapse of galactic civilization, or the deliberate, xenophobic policies of the colony itself). Or it may have been forgotten and it is then (usually) rediscovered. Such things are especially likely to happen in the fifth and sixth steps of Donald A. Wollheim's own **cosmogony of the future** (from his book *The Universe Makers*)—"the Decline and Fall of the Galactic Empire" and "the Interregnum" (43–44). There are lost colonies in the *Stargate* franchise and also in *Star Trek*. Howard Weinstein's *The Better Man* (1994) extends the genre into *Star Trek* novels. There are a great many other literary examples, including Anne McCaffrey's Pern series, Marion Zimmer Bradley's *Darkover*, and Lois McMaster Bujold's Vorkosigan books.

It is fairly common for Earth to turn out to be a lost colony. This occurs in Larry Niven's *Protector*, Ursula K. Le Guin's Hainish series, and Gordon R. Dickson's *Wolfling*. Lost colony stories are similar in many ways to and may be descended from *lost world* stories.

Lost World Story

The discovery of a group of people (sometimes therefore called a lost race story) or an ecosystem previously unknown to what we define as civilization. Thomas More's *Utopia* and Jonathan Swift's *Gulliver's Travels* satirize such narratives, and some of the more famous examples include Jules Verne's *A Journey to the Center of the Earth*, Rudyard Kipling's "The Man Who Would Be King," H. Rider Haggard's *She* and *King Solomon's Mines*, and James Hilton's *Lost Horizon*.

Arthur Conan Doyle's *The Lost World* (which, as its title suggests, played deliberately into the genre) has been influential far beyond the movies and the Australian television series which were made from it. Other writers have been inspired by or have borrowed from Conan Doyle's novel, including Russian geologist and early SF writer Vladimir Obruchev in his 1915 *Plutonia* and in *The Sannikov Land* (1926), Greg Bear's 1998 *Dinosaur Summer*, set in Conan Doyle's *Lost World*, and, of course, Michael Crichton's 1995 novel of the same name, which has two characters referring to the paleontologist John Roxton.

Although such stories have been written less and less frequently since the nineteenth century (when there were still occasional parallels in real exploration and archeology), an astonishing number of movies have carried on the tradition, often (but by no means always) by filming the stories I have just mentioned or making movies that could be considered sequels in some sense, such as Steven Spielberg's *Jurassic Park* films, based on Crichton's novels. The Atlantis legend (which can be traced at least as far back as Plato) is, of course, one of the staples of lost world books and films, as are King Kong and Tarzan movies.

J. R. R. Tolkien's *The Lord of the Rings* (which was heavily influenced by H. Rider Haggard) uses not only Atlantis (in the form of Numenor), but also a lost world (complete with surviving pterosaurs which the Nazgul ride, possibly suggested by the pterodactyls from Conan Doyle's novel) and within that world many separate cultures that have lost track of each other and have even, in some cases, lost parts of themselves, the most obvious examples of this last being Gondor's lost royal line and the Ents' lost wives.

SF writers have written *lost colony* stories and have, in addition, frequently written about literal lost worlds, including Earth itself. One of the more famous stories of the search for Humanity's lost home world is to be found in Isaac Asimov's Foundation series. On television, perhaps the best known example is *Battlestar Galactica* (in both of its incarnations), though the *Star Trek* franchise and *Babylon 5* (which was influenced, according to *TNT's Guide to* Babylon 5, by E. E. Smith's Lensman books, *Lord of the Rings*, and Asimov's Foundation Trilogy) contain what might be considered lost race narratives. See **Cosmogony of the Future** and **Wolfling**.

Monomyth (also Journey of the Hero)

The underlying pattern behind innumerable journeys by individual heroes. Carl Jung says, "The hero figure is an *archetype*, which has existed since time immemorial" (*Man and His Symbols* 61). It is supposed by some mythographers and psychoanalysts to parallel the development of most humans. Otto Rank discussed his version of the myth in *The Myth of the Birth of the Hero* (1914). For him, it was almost entirely bound up with the ferocious relationships of the nuclear family.

James Joyce rather offhandedly created the term "monomyth" in *Finnegans Wake* in 1939, "And then and too the trivials! And their bivouac! And his monomyth! Ah ho! Say no more about it! I'm sorry!" (581). Robert Martin Adams claims (with some justification) that "Reading the *Wake* properly is ... a superhuman task" (33). It's not accidental that Joseph Campbell, the mythographer and scholar who was to make the most important contributions to the idea of the hero's journey in the twentieth century was the author of *A Skeleton Key to Finnegans Wake* and the editor of *The Portable Jung*.

In *The Hero with a Thousand Faces*, Campbell wrote, "The standard path of the mythological adventure of the hero is a magnification of the formula represented in the rites of passage: *separation—initiation—return*: which might be named the nuclear unit of the monomyth" (30). Campbell's version of the hero's journey is more complex than most others, including stages that fit religious leaders as well as the usual figures from myth.

The stages of the hero's journey start before birth (the parents must be extraordinary—often but not always gods or royalty) and end a little after death (when some heroes reach paradise), though each writer has a different description of just what those stages are, and not every hero conforms to the full pattern, whatever that is considered to be. David Adams Leeming says, "The monomyth itself is an expression of the journey of the hero figure and a reflection of our own journey from birth to the unknown.... The hero does what we would all like to do; he literally 'finds himself'" (6).

The Jungian term for such a thing is individuation "the psychodynamic process by which the Self is realized" (O'Neill 175). That is, for example, exactly what Seven of Nine must do in *Star Trek: Voyager*, or as the title of *Star Trek Scriptbooks Book Two* puts it *Becoming Human: The Seven of Nine Saga*. (See **Cassandras**.)

There is a clear indication of the connection between discovering one's humanness and a heroic journey in Poul Anderson's "Time Patrol." The Time Patrol puts Manse Everard through a series of very sophisticated psychological tests to see if he is suitable for their dangerous and demanding job. As he works through them (and by no means coincidentally), he makes an interesting discovery, "He had known it all, of course, but only as isolated shards of fact. It was peculiar, this sudden sensing of himself as an integrated organism, this realization that each characteristic was a single inevitable facet of an overall pattern" (3).

In other words, the hero's journey and the discovery of humanness are complementary activities, and success at one indicates the high probability of success at the other. As Dard Kelm puts it in his casual welcome to the Time Patrol Academy, "We don't need to punish failure in studies, because the preliminary tests have guaranteed there won't be any and made the chance of failure on the job small" (3).

The journey of the hero has become one of the most important plot patterns in science fiction, fantasy literature and elsewhere, and it can be found in material as widely separated in time and intention as *The Odyssey* and *Rocky III*. The plot of *The Lord of the Rings* is, in some ways, a series of variations on the journey of the hero, and George Lucas (who was much influenced by both Joseph Campbell and J. R. R. Tolkien) followed it in his *Star Wars* films.

In his delightfully detailed *The Monomyth in American Science Fiction Films*, Donald Palumbo says, "The monomyth's appearance in each of the first ten *Star Trek* films is worthy of study not only in itself but also precisely because these films share this characteristic with the *Star Wars* films: Both film series are products of the two far larger, multimedia SF entertainment franchises ... that are most obviously in competition with one another as unprecedentedly successful ... cultural phenomena; thus, both may owe much of the impact they have had on the popular psyche to their similar uses of this archetypal material" (142). Palumbo's examples also include (among others) *The Time Machine, Logan's Run, Escape from New York, Dune, The Terminator, Total Recall, The Matrix* and *Back to the Future*.

Here is my version of the monomyth or the stages in the journey of the hero. It is different from, shorter than and perhaps less spiritual than Campbell's, but I've been explaining it successfully to students for years.

(1) THE HERO IS BORN to parents who are divine, royal or extraordinary in some other way.

(2) THE HERO'S LIFE is threatened, and he or she is hidden away, usually in the care of a wise, elderly teacher with magical or divine powers.

Heroes begin life with an aura of power around them. There are marvelous signs that herald their births; their parents are specially important in some way; and heroes may be capable of doing wonders before ordinary mortals are able to do anything. But an aura of power does not promise bright peace; more often, it threatens dark strife and a darker death. For their own protection, heroes are hidden away, usually in the care of wise, elderly, magical teachers, who serve, save and guide them until they have grown great enough to prevail alone. The hiding away of the hero is usually connected to the parents' great position and, on occasion, even greater enemies, but not all heroes start as aristocrats, no matter how they see themselves.

The wise, elderly teacher is most often a character reminiscent of Merlin, such as Gandalf or Obi-Wan or Dumbledore, but in *Star Trek* (2009), Kirk's mentor (the word comes from the disguise Athena assumed so she could advise Odysseus' son when he was in peril) is Christopher Pike, a starship captain who convinces the young troublemaker to join Starfleet and protects him afterward. For Sarah Connor in *Terminator Genisys*, her "Guardian," whom she calls "Pops" is a terminator from the future who has been protecting her since she was nine.

Like the hero, children see themselves as miraculous because each of them is the center of his or her own universe. Small children feel threatened by large forces they do not understand, and see their parents through mists of glory. Their teachers seem impossibly old, remarkably wise and possibly magical—at least until the children grow to understand them and themselves better. One of the more common childhood fantasies is the belief (which comes with disillusionment) that they have been stolen away from true parents and left in the care of people to whom they could not possibly be related.

Sometimes such children go so far as to search for the real parents who gave them up for adoption. In Otto Rank's words, "The entire endeavor to replace the real father by a more distinguished one is merely the expression of the child's longing for the vanished happy time, when his father still appeared to be the strongest and greatest man, and the mother seemed the dearest and most beautiful woman" (67). Or as Witter Bynner put it in his poem "Heroes," "I was a child bereaved of childhood./ Even my closest, even they/ Who loved me, had lessened into faulty clay."

(3) THE HERO PASSES a test which signals the end of childhood.

The hero passes a test which signals the crossing of a threshold. It may be an achievement that is longed for, planned carefully and accomplished with delight, as when Theseus moves the immense stone that conceals his father's sword. Or it may be half-destiny, half-accident, as when Arthur stumbles to the sword in the anvil in the stone and pulls out the blade that burdens him with a kingdom. But always, passing the test brings more problems, complexities and adventures than the young hero had dared to dream, feared to find or imagined to be possible.

Like the hero, we pass our own tests that signal childhood's end. By design or by chance, we find ourselves in a larger world with darker perils and brighter hopes. Our transitions are usually simpler—going to college to get an education and qualify for a place in the world, joining the military before or after college, going to work or getting married. One of my female students came from a Scottish family where everyone was given a sword on her or his eighteenth birthday. It's certainly a transition with a long history in the world and one that may become more popular once light sabers are available on Amazon. All such actions signal to society that we have put away childish things or at least stopped living in our parents' basements and started living in cheap apartments instead. But somehow things never do turn out quite as we had expected that they would.

(4) THE HERO ACCEPTS the quest (usually with reluctance).

Most heroes are stuck with a quest— a war to fight, a kingdom to win or a Golden Fleece to find. Some of them, like Theseus and Beowulf (and most of the captains in Starfleet), can't wait to start killing monsters and winning fame, though James Kirk seems to be wondering about the reason for repeated adventures at the beginning of *Star Trek Beyond* (2016).

The more sensible heroes (and they're the majority) have to be pushed, prodded and cajoled into quests. Arthur puts the sword back into the anvil in the stone and hopes that no one noticed; Odysseus pretends to be insane so he won't have to fight in the Trojan War; and Frodo Baggins offers the ring of power to almost everyone he meets who is not absolutely evil. But eager or reluctant, the hero is stuck and the quest goes on.

Very often the quest is about more than glory. It is frequently bound up with the survival of a close friend, a lover, a city or a world. In the Russian fairy tale "Water of Youth, Water of Life, and Water of Death," a warrior princess leads her army against the Tsar to force him to reinstate her husband as his heir (Pilkington and Pilkington 148–154). In "The Feather of Bright Finist the Falcon," a Russian girl marches thousands of miles to rescue her lover and marry him (*ibid.*, 180–187). In *Star Trek: Voyager*, Captain Janeway, a modern Odysseus, must get her ship and crew home from the Delta Quadrant, a farther shore than any Federation vessel had reached before. In the Terminator series, Sarah Connor (and her son, John) must save humanity from destruction by Skynet.

(5) THE HERO CONFRONTS DEATH and becomes more powerful.

Along the way heroes learn the danger of death and come to terms with it. As Captain Kirk says in *The Wrath of Khan*, "How we face death is as important as how we face life." I know it sounds strange to suggest that it's necessary to learn the danger of death, but most people (at least most people under forty) don't have an emotional (as opposed to intellectual) belief in their own deaths. I've asked that question often enough to know it's true. I can also testify to that from my own experience.

I was in an automobile accident where the car I was driving left the freeway, spun 180 degrees, and landed in a field. The investigating officer did not give me a ticket but told me I had done everything right in the circumstances (the tires of the borrowed car were to blame). His praise for my driving had to do with the fact that in 90 percent of such accidents, the car rolled and the passengers died. It hadn't occurred to me for a second that I was in mortal danger. I simply did not believe in the possibility of my own death. I have learned better since.

The confrontation with death may take many forms, from a straightforward situation that threatens the hero's life to a full-scale journey to the underworld. Odysseus, of course, is the most famous voyager in that realm, but Frodo Baggins' trek through Mordor is a parallel, as are Captain Kate Janeway's struggles in "The Year of Hell," Spock's loss of his world and his mother to a black hole in the 2009 *Star Trek*, and Luke

Skywalker's confrontation with the Death Star. The journey to the underworld counts as a kind of death in and of itself, but it is not unusual for the hero to die and be brought back to life in some fashion, as Kirk is at the end of *Star Trek Into Darkness* and Janeway is at the end of "The Year of Hell." She is restored to life (or some version of her is) with the restoration of the timeline after she destroys the Timeship. (See **Cassandras**.)

Another very common confrontation with death for the hero is the death of a mentor, parent, or foster parent. This happens to (among many others) Frodo, Luke Skywalker, Harry Potter, James Kirk (in the 2009 film), John Connor (in *Terminator II*), Rocky (in *Rocky III*), and to Alexandra (Alex) Owens (in *Flashdance*).

In the end, the hero meets the hurts and hopes of humanness and learns to live with them. What the hero finds is a more powerful, more complete self than the one he or she set out with. We all have our quests—searches we can't avoid and selves we must explore. We must come to terms with our own mortality and humanity. In that sense we are all heroes because we are all human.

(6) THE HERO COMPLETES the quest and achieves glory.

(7) THE HERO LIVES UNHAPPILY ever after.

(8) THE HERO REACHES PARADISE.

The quest is achieved, and the hero gains the glory he or she has expected or dreaded. But glory glows for a moment only and then is gone. Somehow things never move smoothly to their well-planned, well-earned conclusion. Jason doesn't get the kingdom he has spanned the world for; Odysseus reaches the end of his weary wandering only to find a bloody battle with fools and another quest beyond that. The hero journeys not in the romance of happily ever after but in the real world where struggles don't stop while life lasts and where heroes live unhappily ever after, or at least until they pass into the strange dark of death or the bright dawn of paradise.

Some heroes, like King Arthur or Frodo or James Kirk live unhappy lives followed by a passage to Avalon or the Undying Lands or the Nexus. As might be expected, Kirk is the only one to escape from an idyll to a deadly struggle, though Daniel Jackson (from the *Stargate* franchise), having achieved apotheosis, gets himself kicked out of a transcendent existence. It is definitely fair to say that any heroic quest that comes with a sequel cannot end happily. If P. J. Travers' Banks family had lived in unmitigated bliss, Mary Poppins could never have returned. In that sense, life is a series of sequels.

Like the hero, we find that no quest accomplished or glory earned protects us from the burdens of humanness. The troubles never disappear, the triumphs never vanish while we live with ourselves in the real world. Sometimes, though, we may find that we have company on our hero's journey. As Donald Palumbo says, "In a sense the recurring monomythic hero in the *Star Trek* films is the *Enterprise* crew as an ensemble and, by extension, the ship itself" (142). In the words of Witter Bynner's poem "Heroes," "We are all such little heroes in the sun." Perhaps the journey of the hero is a path we all travel together in storm, in shadow and finally for a time at least, in sunlight.

Proto-Science Fiction

Much early fiction contains elements which have found their way into science fiction. In Book 18 of Homer's *Iliad*, Hephaestus (Vulcan) is helped and supported by robot maidens who "were like real young women, with sense and reason" but made of gold (293). (See **Robot**.) Book VIII of Homer's *Odyssey* has ships which are sentient and telepathic. King Alcinous says to Odysseus, "Tell me ... your city, that our ships may aim at the right place in their minds. For we Phaiacians have no pilots; our ships have no steering-gear ... but they understand of themselves the thoughts and intentions of the seamen; they know all the cities and countries in the world, and cross the gulf of the sea at full speed" (Rouse 99).

Other examples include Lucian's *True History*, Dante's *Divine Comedy*, Shakespeare's *The Tempest*, Cyrano de Bergerac's *Voyages to the Moon and the Sun*, and Voltaire's *Micromégas*. Robert Silverberg would start the list even earlier than Homer. He says in the Introduction to his novel *Gilgamesh the King*, "The Gilgamesh epic ... must be the earliest science-fiction story still in existence, for surely the tale of a quest for an immortality serum qualifies as science fiction."

For some this is merely an indication that science fiction has always existed, but for others, these works are forerunners, and the true genre doesn't emerge until the time and scientific understanding are ripe. One argument is that SF cannot exist until the speed of change is so great that it can be clearly observed in a person's own lifetime. Of course, for us now it can sometimes be clearly observed in a person's own weekend. Jeff Prucher says the first use of the term "proto–science fiction" was by Damon Knight in 1962 (157). See my Introduction and the **science fiction** entry.

Rim World (or Rimworld)

A world far from the center of the galaxy and equally far, the assumption goes, from galactic civilization. The term was first used by Robert A. Heinlein in *Citizen of the Galaxy* in 1957, "H.G.C. *Hydra* lifted from Hekate, bound for the Rim worlds." Earlier in the same novel, he had said, "Free Trader *Romany Lass*, bound for the Rim."

The argument is that galactic empires, such as the one in Asimov's *Foundation* series, tend to move from maximum concentration at the center of the galaxy to thinly populated or nonexistent at the rim; hence, rim worlds become the frontier in galactic civilizations, where frontier "marshals" (like Captain Kirk) enforce the law with phasers. Heinlein says, "*Hydra* was cruising above speed-of-light toward the Rim world Ultima Thule, where she would refuel and start prowling for outlaws."

There is a similar dynamic but with the heroes on the other side of the law in the Joss Whedon series *Serenity*. In *Star Wars: The Force Awakens*, when Finn is thinking of running away, Maz Kanata tells him, "You see those two? They'll trade work for transportation to the Outer Rim."

For novels of this type see A. Bertram Chandler's *Rim Worlds* books, C. J. Cherryh's *Rimrunners*, and Gordon R. Dickson's *Wolfling*. While this emphasis on the frontier explains part of the appeal of such stories, there is an important additional element: On a star map that includes Earth, the You-are-here arrow points to the rim. Or as Carl Sagan put it in *Cosmos*, "We find that we live on an insignificant planet of a humdrum star lost between two spiral arms in the outskirts of a galaxy which is a member of a sparse cluster of galaxies, tucked away in some forgotten corner of a universe in which there are far more galaxies than people" (193).

Comments about the place of humanity in the universe aside, there is an example here of a common problem in science fiction: believing that what happens on Earth will find a ready parallel in outer space. Thus, galactic empires will be our own imperial histories writ large, aliens will share our motivations, both good and bad, and nations and galaxies will inevitably be organized in the same fashion, even though there are very seldom super-massive black holes occupying the main population centers of nations.

Of course, science fiction is not exclusively about space, the future or technologies we haven't invented yet. Often, it is a mechanism that allows us to say what our own societies would otherwise prevent us from saying. Isaac Asimov's robots were, among other things, an attack on racism (as Mr. Spock was in the original *Star Trek*). And in such cases, the parallels are examples of deliberation, not carelessness. We also, as David Brin suggests in the Foreword to this book, use SF frequently and deliberately to explore history—and the other way round. (See **Wolfling**.)

Ruritania

An imaginary, German-speaking country somewhere in the center of Europe created by Anthony Hope in his novels *The Prisoner of Zenda* (1894) and *Rupert of Henzau* (1898) and hence any similarly nostalgic, aristocratic, politically labyrinthine story. In film, the 1956 Grace Kelly vehicle *The Swan* (based on the Molnár play), Laurence Olivier's *The Prince and the Showgirl* (1957) and Peter Ustinov's *Romanoff and Juliet* (with added material from Shakespeare's play and Cold War brinksmanship, 1960) have all moved the subgenre from an emphasis on adventure to romantic comedy.

The dangers inherent in international politics (as opposed to local mayhem) are indicated by the response from the President of Concordia in *Romanoff and Juliet* when he's asked where his country is, "Why should I tell you? The reason it still survives may be because no one can find it."

Robert A. Heinlein's *Double Star* (1956) and Edmond Hamilton's *Star Kings* (1949) rework the plot of *The Prisoner of Zenda*, and SF has found room for a number of fairy tale, middle European kingdoms (or their outer space counterparts), complete with corkscrew politics. In fact, *The Encyclopedia of Science Fiction* says, "The palace-politics which govern many Galactic Empires owe more to Hope than they do to Edward Gibbon" (Clute and Nicholls 1034). Isaac Asimov claimed to have based his Foundation series on Gibbon's *Decline and Fall of the Roman Empire*, but even his politics sometimes seem more Ruritanian than Roman.

Charles Sheffield is possibly glancing at one of his own minor inspirations when, in *Cold as Ice*, he has Undersecretary Posada ask Jon Perry, "And do you usually dress like the head pimp of the Ruritanian navy?" (85). However, such references are common and often completely facetious, playing high officers in the naval forces of a small, landlocked country for laughs, as P. G. Wodehouse does in *Spring Fever* (1948), "On the sidewalk outside the main entrance of Barribault's Hotel there is posted a zealous functionary about eight feet in height, dressed in what appears to be the uniform of an admiral in the Ruritanian navy" (177).

Ruritanian elements often appear in **lost colony** and **lost world** stories. A wide variety of authors from Edgar Rice Burroughs to Anne McCaffrey have used Ruritanian elements in their work, but perhaps the most successful recent series to employ them is Lois McMaster Bujold's saga of Barrayar and the adventures of Miles Vorkosigan. (Here too the dangers of politics on a larger scale—interstellar this time—are far greater than the local variety. Barrayarans are nearly exterminated once their planet is found.)

The Encyclopedia of Science Fiction points out that such "tales set in Balkan enclaves should be called Graustarkian, after the otherwise very similar *Graustark* (1902) by the US writer George Barr McCutcheon" but declares such a procedure "pedantic and unproductive" (Clute and Nicholls 1034).

What's really going on here is a distinction between imaginary German-speaking countries and imaginary Slavic-speaking countries, which, given the fact that the two groups squared off in two world wars, might not be so pedantic after all. Another distinction is between Central Europe for Ruritania and Eastern Europe (rather than the more restrictive Balkans) for Graustark.

A good example of the difference is the 1997 film *The Beautician and the Beast*, starring Timothy Dalton and Fran Drescher. It is clearly Graustarkian, set as it is in an imaginary country called Slovetzia on the northern border of Romania with a ruler who is described as "Stalin without the charm." The same might indeed be said for Bujold's Barrayar series, replete, as it is, with Russian cultural elements. Though the Barrayarans don't actually speak Russian, there is a Russian-speaking minority on the planet. Lois McMaster Bujold calls her military aristocracy the "Vor Lords." "Vor" is Russian for thief, which is certainly an appropriate designation for a hereditary aristocracy. There are also references to Baba Yaga (the most famous witch in Slavic fairy tales).

In addition, Bujold has included a character in the person of Lord Ivan Vorpatril who is immediately recognizable from Russian fairy tales, novels and even movies. He is Ivan the Fool (known in the Vorkosigan books as "that idiot Ivan"), a feckless, sometimes reckless, but always kindhearted and good-natured individual who blunders inevitably into danger and stumbles miraculously out of it. What's the point of genre definitions if we can't enjoy a little pedantry?

Science Fantasy

The *OED* identifies two meanings for the term. "A genre of fiction that combines elements of science fiction and fantasy" (1948). Also: "Science fiction which is characterized by the depiction of technologies generally regarded as scientifically impossible" (1951). Science fantasy may include alternate worlds, monsters, parapsychological powers, or even mythology, magic and so forth, usually within a technological context. The fantasy elements may be rationally explained in some pseudoscientific fashion.

Authors who have written science fantasy include Marion Zimmer Bradley, Edgar Rice Burroughs, Anne McCaffrey, C. L. Moore, and Gene Wolfe, to mention only a few. George Mann maintains that "at least part of the success of Frank Herbert's *Dune* (1965) can be put down to his ability to blend the feel of a fantasy saga with what is ... an explicit Space Opera" (506). He continues, "The movie *Star Wars* and its sequels can also be viewed as science fantasy" (507). Another example is the J. J. Abrams television series *Alias* (2001–2006).

Indeed, it is this mixture of things that can result in a seemingly strange success, giving SF a larger canvas than even that immense genre sometimes has and providing for fantasy a more believable habitation than we have come to expect.

Science Fiction

The *OED* finds an isolated example (by W. Wilson) as early as 1851, though it has the meaning of fiction about science. The term as we now use it is credited to Hugo Gernsback, who employed it frequently in his magazine. The *OED* indeed cites the first use of the term in *Amazing* in 1927.

Most definitions of SF are argumentative at best and abusive at worst and can often be boiled down to the message that what I write (or read) is science fiction, while what you write (or read) is some sort of amorphous glop left over from a particularly repulsive evolutionary dead end.

Isaac Asimov's definition seems to me to be as sensible as any (and far more sensible than most). He says (in *Asimov on Science Fiction*) that SF is about "events played against social backgrounds that do not exist today" but "could, conceivably, be derived from our own by appropriate changes in the level of science and technology." He goes on to argue "that the field can scarcely have existed in its true sense until the time came when the concept of social change through alterations in the level of science and technology had been evolved in the first place" (17–18). He indicates the Industrial Revolution as the watershed event and declares, "Science fiction had to be born sometime after 1800" (18).

It is, of course, possible (in the nature of arguments about science fiction, it may almost be mandatory) to make different assumptions. Adam Roberts, in his *The History of Science Fiction*, declares emphatically, "Science fiction was reborn in one year, 1600, the year that the Catholic Inquisition burned Giordano Bruno ... for arguing ... that the universe was infinite and contained innumerable worlds ... a fundamentally science fictional conception" (36).

Brian Stableford begins the Chronology of his *Historical Dictionary of Science Fiction Literature* in 1726, and cites for 1771, "*L'An 2240* by Louis-Sebastian Mercier, the best-selling book of its era in France" and "the first.... Utopian society situated in the future whose evolution has been enabled by technology" (xiii).

The simplest definition of SF is that it is about social change and an extrapolation from the science of the present to a possible shape for the future. In the words of Jules Verne, "I merely use my imagination and literary skill to argue from what is possible to what may be possible—tomorrow" (Taves and Michaluk 53). However, there are a few additional elements. As part of its concern for the nature and passage of time, science fiction also frequently explores both the past and possible parallel worlds. And finally, SF's explorations go beyond human society and Earth to the universe and the other intelligences that may exist somewhere in it.

This definition would include American genre SF, much that was written in the nineteenth century (including Mary Shelley's *Frankenstein*), and the British scientific romances (plus, of course, Jules Verne), but it would exclude everything before the scientific method became a commonplace and the "rate of change, and the extent of the effect of that change on society" became "great enough to be detected in the space of an individual lifetime" (Asimov *On Science Fiction* 18).

I would not look for the true genre as early as Adam Roberts does, but I would call a number of earlier works going back at least as far as Homer and including Shakespeare's *Tempest* and Cyrano de Bergerac's *Voyages to the Moon and the Sun* **proto-science fiction**. They sit on the terminator separating the benighted past from the well-lighted present but are already edging out of the darkness. And if that isn't a sufficiently grandiose view of the SF field, I don't know what is.

Charles Sheffield also found metaphor helpful in his definition, "Science fiction consists of stories set on the shore or out in the shallow coastal water of that huge scientific land mass. Stay inland, safe above high tide, and your story will be not science fiction, but fiction about science. Stray too far, out of sight of land, and you are in danger of writing fantasy—even if you think it's science fiction" (*The Borderlands of Science* 7).

Coined in 2001 by Jennifer Rohn, Lab Lit is a comparatively recent genre name for those works that "Stay inland, safe above high tide." They are stories, she writes, "in which scientific characters, activities or themes are portrayed in a realistic manner," though even they can sometimes go too close to the shore and become SF, an understandable danger when science and science fiction intertwine so effortlessly.

Incidentally, for those who don't like Asimov's definition or mine or Charles Sheffield's, here is the *OED*'s: "Fiction in which the setting and story feature hypothetical scientific or technological advances, the existence of alien life, space or time travel, etc., *esp.* such fiction set in the future, or an imagined alternative universe."

Here's one more, the shortest yet, from *The Star Trek Encyclopedia: A Reference Guide to the Future*, "Branch of literature and media dealing with the effects of science and technology on society" (Okuda, Okuda, and Mirek 658). For even more definitions see **Speculative Fiction** and **Speculative History**. (Some parts of this entry appeared in my Introduction to *The Fantastic Made Visible*.)

Scientific Romance

The main term used in Britain to describe science fiction up until the mid 1940s or, if George Mann is correct, as late as the 1960s, with "some nostalgic examples ... produced today" (507). Though, as usual with genre terms, there were many others employed and a fair amount of disagreement. Jules Verne's novels were generally described as scientific romances in Britain, but the French is *roman scientifique* or scientific novel.

The OED finds the first use of the term in 1797 by Isaac D'Israeli (father of British Prime Minister Benjamin D'Israeli) and defines it as "a work of speculative fiction or science fiction, esp. one characterized by a spirit of serious intellectual inquiry." "In 1866 the *Nation* described Oliver Wendell Holmes' *Elsie Venner* as 'a scientific romance of the destiny which a physican can discern in blood and nervous tissue'" (Nevins and Martinez. For more on *Elsie Venner*, see **Mutant**.)

The *Encyclopedia of Science Fiction* maintains that scientific romances are "characterized by long evolutionary perspectives; by an absence of much sense of the frontier and a scarcity of the kind of Pulp-Magazine-derived hero who is designed to penetrate any frontier available; and in general by a tone moderately less hopeful about the future than that typical of genre sf until recent decades" (Clute and Nicholls 1076).

Roberta Rogow's *Futurespeak*, on the other hand, defines science romance as "A story that deals largely with the betterment of society through technology as depicted in the works of Jules Verne and his followers" (298). There is nothing contradictory about these two definitions, provided we remember that the first applies primarily to H. G. Wells, Arthur Conan Doyle, and so on, and the second to Jules Verne.

Brian Stableford adds, "The most conspicuous difference between scientific romance and science fiction ... was the total absence from scientific romance of the myth of the Space Age. Although scientific romances did feature expeditions into space, they never portrayed such expeditions as the initiation of an inexorable sequence of colonial expansion" (*Science Fact and Science Fiction* 469). (See **Cosmogony of the Future**.)

H. G. Wells was not entirely happy with the term. He puts it disparagingly into the mouth of one of his characters in his 1923 novel *Men Like Gods*, "He thinks that things that don't exist *can* exist. And now he imagines himself in some sort of scientific romance and out of our world altogether." However, lest we take the term too seriously, I point out that in 1934, in a Preface for a collection of seven of his novels published by Alfred A. Knopf, he wrote, "For some years I produced one of these 'scientific fantasies,' as they were called, every year" (v). He also used the phrases "for the writer of fantastic stories" and "the fantasy writer" (iv). In addition, he declared, "These tales have been compared with the work of Jules Verne and there was a disposition ... at one time to call me the English Jules Verne. As a matter of fact there is no literary resemblance whatever between the anticipatory inventions of the great Frenchman and these fantasies." He is making a distinction which is at least partly valid. Wells continued, "Many of his inventions have 'come true.' But these stories of mine collected here do not pretend to deal with possible things; they are exercises of the imagination in quite a different field" (iii).

It would be hard to find a better example of the difficulties of labeling any genre. This is especially true since H. G. Wells is quite right in what he says, and on some level, he is indicating very clearly the difference between British works of the time and what Jules Verne had been doing considerably earlier. But the issue becomes even more complicated because while Wells was right, he also turned out to be wrong. His stories, though they may not have been meant to deal with possible things, succeeded sometimes in predicting and even in helping to bring about scientific advancements. (See **Atomics**.)

As Donald A. Wollheim summed it up in *The Universe Makers*, "Wells was a Utopian and though his writings were to encompass as many Remarkable Inventions as Verne—indeed to help fill out the armory of modern science fiction—it was his Social Satire that was dearest to his heart and was the lever with which he sought to move the world" (20).

Sci-Fi (also Sci Fi)

There is an argument about this term, and of course, arguments about genre terms are the rule rather than the exception. For years, I told my *science fiction* classes, "This is not a synonym for or abbreviation of science fiction. And then to prove my point I quoted Isaac Asimov on the subject: "'Sci-fi' is now widely used by people who don't read science fiction. It is used particularly by people who work in movies and television. This makes it, perhaps, a useful term. We can define 'sci-fi' as trashy material sometimes confused, by ignorant people, with s.f. Thus, *Star Trek* is s.f. while *Godzilla Meets Mothra* is sci-fi" (*Asimov on Science Fiction* 27).

It's hard to find a more authoritative voice in science fiction than Asimov, but he was certainly not alone. In her book *Futurespeak: A Fan's Guide to the Language of Science Fiction*, Roberta Rogow defined it as, "Abbreviation for Science Fiction first used by Forrest Ackerman in the 1930s. Most Mundanes (including many book and film critics, alas) persist in using it. Truefen prefer SF (which is the form used in this book)" (299). "Mundanes" are people who are not directly connected with science fiction, and "Truefen" are true fans.

Clearly, this is, in part, a question of us versus them. The *OED* reflects that in its first definition, "SF is the preferred term for some science fiction enthusiasts, for whom *sci-fi* carries a connotation of poor quality or unoriginality." They cite the *Observer* from 1974, "The SF fan world abounds in language ... that can baffle the novice.... Most important of all, you must not say 'sci fi'—it's always SF."

However, times, attitudes and words change. Asimov's book was published in 1981, and Rogow's in 1991. In *Insistence of Vision*, a 2016 collection of stories and essays, David Brin writes about what he calls "The Heresy of Science Fiction," using Sci Fi, science fiction, and SF interchangeably (16–23). (He also argues that a better name for the field would be *speculative history*. See that entry and also David Brin's Foreword to this book.) And David Brin is not alone in his position any more than Asimov was alone in his.

Bergman and Lambert in their *Geektionary* (2011) use "hard sci-fi" (29), which is perhaps the ultimate sign of approval since **hard sf** is considered by many to be the purest expression of science fiction. Grazier and Cass chronicle the change in *Hollyweird Science* (2015), "The historical struggle for respectability gave rise to various internal divisions.... Consequently, the term 'sci-fi' was used for decades in a mostly pejorative sense.... As the genre's critical reputation has improved, creators and fans have pretty much chilled out about this terminology and today 'sci-fi' is, essentially, synonymous with 'science fiction'" (12). While that's not entirely true, it's reasonably accurate.

Science fiction has become much more successful and, as is the way with such genres, has expanded outward in an astonishing fashion. In a field that includes (to take only one small area) cyberpunk, steampunk, atompunk, nanopunk and biopunk, sci-fi seems to be a term from the good old classical days when John W. Campbell's *Astounding* was on the newsstands and radioactive monsters were in the movies.

Separable Soul

The idea that an individual's life force can exist apart from his or her body, usually hidden away in some unexpected or inaccessible place for safety. This is a frequent motif in folklore, myth, fantasy and science fiction, where the separable soul renders its owner invulnerable and/or immortal as long as the soul's receptacle is not harmed. In the "Whatever Happened to Mr. Garibaldi?" episode of *Babylon 5*, Lorien says to Captain Sheridan, "We all have secrets and surprises. Did you know you have a Vorlon inside you? Well, a piece of one."

In one version of the tale of the Trojan War, Paris's soul resided in a log which was almost burned but saved at the last moment, hence, Marion Zimmer Bradley's title for her novel *The Firebrand*. In Norse myth, the separable soul of Balder, the oak tree god, was supposed to reside in the mistletoe plant. In the Slavic fairy tale "The Dragon and the Prince," the dragon's strength is to be found in another empire in a lake. "In that lake is a dragon, and in the dragon a boar, and in the boar a pigeon" (Pilkington and Pilkington 116). Other fictions that have used separable souls include the Harry Potter series, with Voldemort's multiple hiding places, and *The Lord of the Rings*, with Sauron's One Ring.

Speculative Fiction

The *Collins English Dictionary* defines it as "a broad literary genre encompassing any fiction with supernatural, fantastical, or futuristic elements." But not everyone would agree. It was first used in 1889 and then, more importantly, by Robert A. Heinlein in 1947. His most elaborate definition of the term is in a letter written on March 4, 1949, and published in *Grumbles from the Grave*.

Having first established that the "overworked" part of science fiction "consists of stories about the wonderful machines of the future which will go striding around the universe," Heinlein says, "Speculative fiction (I prefer that term to science fiction) is also concerned with sociology, psychology, esoteric aspects of biology, impact of terrestrial culture on other cultures we may encounter when we conquer space, etc. without end." It's clear, though, that he sees the territory of science fiction and speculative fiction as roughly coterminous. He continues, "However, speculative fiction is not fantasy fiction, as it rules out the use of anything as material which violates established scientific fact, laws of nature.... [I]t must be possible to the universe as we know it" (49).

When he wrote the letter, Heinlein was complaining about an editor who had rejected *Red Mars*. It was his way of explaining the real nature of the stories he wrote. Some of the New Wave science fiction writers of the 1960s and 1970s used Heinlein's term in order to do the same thing, and like Heinlein, they wanted to make it clear that their genre was larger and more complex than many people thought. As James Gunn put it, "Younger writers picked up or reinvented Heinlein's 1947 term for science fiction, 'speculative fiction' on the grounds that 'science fiction' is too narrow to cover the kinds of fiction that qualified under any reasonable definition but included no science."

Gunn then does a bit of analysis, "Their motivations probably are a bit more complex: the term 'science fiction' is not broad enough to cover the kind of fiction they wanted to write, and a new name suggested new possibilities and new directions—and concealed the shameful pulp origins" (*Alternate Worlds* 236).

There seems to have been (and very likely there still is) a certain amount of concern about what counts as respectable literature. And in the process of broadening the definition of speculative fiction (sometimes called speculative literature) those SF (the initials work for either designation) writers who were specially concerned with certain kinds of style and narrative devices had managed to make a category that included not only science fiction but also postmodernist fiction and magic realism.

In "Margaret Atwood and the Hierarchy of Contempt," Peter Watts writes, "You might expect respectability to correlate with real-world plausibility in the narrative itself. You would be wrong. The same critics who roll their eyes at aliens and warp drive don't seem to have any problems with a woman ascending into heaven while hanging laundry in *One Hundred Years of Solitude*, just so long as Gabriel García Márquez doesn't get published by Tor or Del Rey" (1). As usual with genre criticism and even the terms themselves, almost everything is in the eye of the beholder.

Here is one of Watts' conclusions (though I point out that he has an axe to grind or more likely a beam weapon to calibrate): "Science fiction has become more relevant than 'Literature.' It could hardly be otherwise. Here in the real world, people run software with their brainwaves. Robot dogs are passé. Teleportation is a fact. It has become routine to genetically cross goats with spiders, fish with tomatoes" (3). From my perspective as a professor of literature and history with an Oxford doctorate, I would say that science fiction is literature, and anyone who looks at the new doctoral dissertations that are being produced all over the world would have to agree. Though Jack Williamson may have been unusual when he got a Ph.D. in 1964 for writing about an SF author (H. G. Wells), he has plenty of company now.

More importantly (for this entry at least), the definition for speculative fiction has continued to expand until it includes almost anything with a dash of unreality, including *fantasy*, horror and *alternate history*. The same monster movies that were disparaged as *sci-fi* are now part of that very large category. And more important still, this has been a part of an expansion of science fiction and its related genres, a sort of cultural conquest that is, for many, quite surprising.

I could provide a whole survey of examples, but here is one that may seem relevant in the midst of a literature versus SF discussion. In 2013, the Utah Shakespeare Festival (where I work in the summers as literary seminar director) did a production of *The Tempest*, and I wrote an article about the play for their online magazine. Both of those things had happened before, but this time, I found myself linking Prospero to the genre of fantasy and to other wizards— in movies, television, books and even theme parks. They were everywhere.

J. R. R. Tolkien's Gandalf had just been featured in the first film of a three-part adaptation of *The Hobbit*. The six books of the Harry Potter series were inescapable, and their eight film incarnations (completed in 2011) had racked up $7.7 billion, making it the highest grossing film series ever, if inflation wasn't taken into account ("Movie Franchises"). *Oz the Great and Powerful* was about to be released in 3D, 2D, and digital versions. The BBC television's *The Adventures of Merlin* was running on the Syfy channel in the U.S., and had been broadcast in 182 other countries (Steve Clarke). Disney had recently purchased the *Star Wars* franchise, which meant there would soon be additional films, complete with more versions of George Lucas's wizards in space. And this was without mentioning writers such as Terry Pratchett and Jim Butcher, whose very successful careers had been driven by men with magical wands.

What is extraordinary about this situation is not that fantasy and science fiction are overwhelmingly popular, but rather how thoroughly they have permeated and even transmuted our culture, subsuming and even replacing other forms of literature. In the past, *The Tempest*, as a remarkable classic by the greatest writer in English, would have remained remote from anything even resembling genre fiction. Now, however, the play is freely discussed as fantasy, and in London, Shakespeare's Globe took advantage of the connection by casting Colin Morgan, the star of *Merlin*, as Ariel in their 2013 production.

Even more interestingly, *The Tempest* has been fitted into the category of **proto-science fiction**, early works on the way to what the genre would eventually become or (if we accept the timeline of Adam Roberts, which begins with the death of Giordano Bruno) science fiction itself. Unlikely as it seems, the Bard and the cosmographer could have crossed paths when Bruno lectured "on Copernican cosmology at Oxford in 1583" (Brigden 306).

For their 2013 Prospero, the Utah Shakespeare Festival cast Henry Woronicz, who had played three different species of aliens on *Star Trek: The Next Generation* and *Star Trek: Voyager*. Not surprisingly, *The Tempest* has been repeatedly adapted as SF, including the 1956 film *Forbidden Planet* and the "Requiem for Methuselah" episode of the original *Star Trek*. The utopian/ dystopian elements inspired Aldous Huxley's 1932 novel *Brave New World*, and Joss Whedon's *Serenity*, the 2005 movie spinoff of the SF television series *Firefly*.

In short, even Shakespeare has been colonized, which should not come as a shock because from science fiction to fantasy, from magic to science and back again, humans have always been good at turning dreams to reality, and now we stand at a point where our new realities may transcend even our oldest, wildest imaginings. Our genres must get bigger to match our dreams, or perhaps, just to cope with our new realities. (An earlier version of this entry appeared in my Introduction to *The Fantastic Made Visible*.)

Speculative History

David Brin's alternative term for *science fiction*, suggesting that the most important element in SF is not science or technology but change. Here's part of what he said in the Foreword to this book: "'Science fiction' was always an iffy term.... What name would I prefer? *Speculative History!*"

In "The Heresy of Science Fiction," an essay from his 2016 book *Insistence of Vision*, Brin writes, "I've long felt that SF should have been named *Speculative History*, because it deals most often with thought experiments about that grand epic, the story of us. Sending characters into the past, or exploring alternate ways things might have gone. Or else—most often—pondering how the great drama might *extend further*, into tomorrow's undiscovered country." He concludes, "we can make do with 'science fiction' as the term for what we do. But time remains the core dimension, vastly more important to our stories, our passions, our obsessions, than technology or even outer space" (18).

As usual, Brin is very persuasive, and he has the facts on his side. Harry Turtledove says, "I'm a science-fiction writer and a historian. The combination is not as uncommon as it sounds—to name just a few, Barbara Hambly, Katherine Kurtz, Judith Tarr, Susan Schwartz, and John F. Carr all use what they learned in college to give depth and authenticity to the worlds they create" (*Agent* 5).

Melissa Scott is another science fiction author with a Ph.D. in history. David Weber says, "I had always intended to get my master's and then my doctorate in history" ("Biography"). Science fiction icons Andre Norton, L. Sprague de Camp and Poul Anderson have written historical fiction. De Camp's nonfiction book *The Ancient Engineers* is, as Harry Turtledove says, "a classic in the field" ("Introduction").

Robert Silverberg wrote historical nonfiction, including *Lost Cities and Vanished Civilizations* and *Empires in the Dust*. Isaac Asimov wrote a nonfiction history series for Doubleday, starting with Greece and covering much of Western Europe and America. Hari Seldon, the originator of psychohistory in Asimov's *Foundation* series, is supposed to owe a debt to Edward Gibbon, just as the series itself owes a debt to Gibbon's *Decline and Fall of the Roman Empire*.

Historians themselves have crossed over into SF territory in a series of anthologies and other speculations about **alternate history**. And now and again, historical novelists move from their usual territory into SF as James Michener did in the course of his novel *Space*.

Science fiction really is about a recognition of the nature of time and change. Asimov says of the genre "that the field can scarcely have existed in its true sense until the time came when the concept of social change through alterations in the level of science and technology had been evolved in the first place" (*On Science Fiction* 18). Our shifting attitudes toward time become clear in a variety of ways. "The casual mixing of time periods that was common in the fictions and the histories of the Middle Ages has disappeared. As Lord Acton described it in his "Inaugural Lecture on the Study of History," "The Middle Ages, which possessed good writers of contemporary narrative, were careless and impatient of older fact. They became content to be deceived, to live in a twilight of fiction" (Acton 4). In part, that was because the majority of medieval historians were ecclesiastics who were more interested in sustaining faith than discovering facts. Beginning with the Renaissance, many things were transformed, including attitudes to history. In Lord Acton's words, "The sixteenth century went forth armed for untried experience, and ready to watch with hopefulness a prospect of incalculable change" (4).

Jeremy D. Popkin says in *From Herodotus to H-Net* (2016), "In studying the historians between 1450 and 1800, we are examining the origins of many of the features of history as it is researched and presented today" (48). The Renaissance produced a new awareness of time periods, an understanding of the changes in languages over time (a great help in evaluating manuscripts), and also the beginning of archeological study. History was popular. In London, "There were two ways of learning about national history in Shakespeare's time, if you did not want to resort to books: you could go to Westminster Abbey, pay a penny and be instructed about the 'living monuments' of dead kings or you could go to the theatre, pay a penny and see the great kings stride out in front of you" (MacGregor 89). (See **Anachronism**.)

Today, we strive to understand not only the facts of the past, but also the nature of past times and how people thought and felt centuries ago. In our post–Einstein world we fixate on the passage of time, thinking of it as the timestream and often visualizing it as a river sweeping onward and branching off into alternative histories or parallel worlds. (See **Time Machine**.) And even in our future-oriented society, we obsess and fight over history, for as Shakespeare said in *The Tempest*, "The past is prologue," and more than ever before, we have come to view the present and the future as parts of, and as the outcomes and consequences for the past.

Speculative Nonfiction

The use of the imagination to explain, describe or predict within the world of facts rather than fiction, including insights and even hypotheses in the sciences and other scholarly disciplines. Such uses of the imagination go back to the very beginning of human thought—to the history of Herodotus or the dialogues of Plato. Indeed, the beginning place may be the beginning of humanness.

I would point to the words of the American anthropologist and historian of science Loren Eiseley, who argues that it was imagination that first made us human, Man "was becoming something the world had never seen before— a dream animal—living at least partially within a secret universe of his own creation.... The unseen gods, the powers behind the world of phenomenal appearance, began to stalk through his dreams" (*Immense* 120). Or as Shakespeare puts it in *The Tempest*, "We are such stuff as dreams are made on."

Brian Stableford says, "Speculation

plays a key role in scientific thought, which overlaps that of extrapolation but usually makes more use of imagination in proceeding far beyond matters of logical deduction. It is the process used to generate hypotheses, including specific predictions, whose subsequent testing by observation and experiment produces and refines scientific knowledge" (*Science Fact and Science Fiction* 497).

Clearly the mechanism of speculation is one of the junctures between science and science fiction. Imagination plays over the same materials and often produces complementary results or results that are illuminating to scientists, science fiction writers and that third group, itself a kind of overlap, futurists. As Stableford points out, "It is not surprising that some of the most successful popularizers of twentieth century science—notably Isaac Asimov and Arthur C. Clarke—were equally successful in the fields of speculative fiction and nonfiction, routinely transferring skills between their short stories and their essays" (498). As this book demonstrates, imagination is the common coin between science and science fiction, and perhaps speculative nonfiction might be called common ground.

Technothriller

The term was first used in 1986 according to the *OED*, but the thing itself was in existence much earlier, with Jules Verne as one of its first creators. The novels of Tom Clancy (and the movies made from them) fall into this category as do most spy stories, including the Jason Bourne novels and films and even more spectacularly, the longest continually running (and most profitable, if the revenue is adjusted for inflation) series in movie history, the James Bond films (26 so far) and, of course, the novels behind them.

A technothriller is a story set in the very near future (or even the present) with technological elements that may or may not be science fiction; it is often difficult (and often deliberately made to be difficult) to tell. A good example that does not involve spies is *Jurassic Park* and its sequels (again in film and print). At least a part of the appeal of such novels and movies is the rapidly shifting line between fiction and fact created by the speed of technological change (Pilkington "Introduction" *The Fantastic Made Visible* 2).

To take a comparatively unbelievable instance which comes from what should be (but is not entirely) a fantasy movie, as I watched the flying aircraft carrier in *Marvel's The Avengers*, I found myself wondering not only if the U.S. military *could* build such a thing, but also if, in some remarkable feat of secrecy, they already *had*. And indeed the response from an official at Naval Air Systems Command to questions such as mine concerned financial and not technological feasibility, "The modern military's budget would likely burst trying to build and operate a full-size flying aircraft carrier that weighs 100,000 tons and stretches the length of three football fields" (Hsu).

Not only do technothrillers explore the cutting edge of scientific transformation, they also, in the nature of dramatic presentations, go over that edge, and frequently criticize its possibly painful outcomes. Like the *sci-fi* movies of the 1950s, which were haunted by radioactive monsters and alien invasions, technothrillers simultaneously explore the present, extrapolate the future and compartmentalize the actual and hypothetical dangers, placing them neatly in a format that allows for their neutralization in the time it takes to watch a movie or read a book. In effect, technothrillers provide both a safety valve, a means of confronting the fears that an unpredictable future inspires, and a safety net, the reassurance that there will be in these scientifically infused tales a good outcome that will keep us safe in our imaginary worlds and in our real lives.

Works Cited

Print and Internet

Acton, Lord John Dalberg. *Lectures on Modern History*. London: Macmillan, 1921. Print.
Adams, John. "Interview." *Doctor Atomic*. Dir. Peter Sellars. Opus Arte, 2008. DVD/Opera.
Adams, Robert Martin. *After Joyce: Studies in Fiction After Ulysses*. New York: Oxford University Press, 1977. Print.
Adler, Jeremy. "Introduction." *The Life and Opinions of the Tomcat Murr Together with a Fragmentary Biography of Kapellmeister Johannes Kreisler on Random Sheets of Waste Paper Edited by E. T. A. Hoffmann*. E.T.A. Hoffmann. New York: Penguin Classics, 1999. Amazon Kindle.
Allen, Grant. "Pausodyne." *Strange Stories*. Seattle: Amazon Digital Services, 2012. Amazon Kindle.
Amis, Kingsley. *New Maps of Hell: A Survey of Science Fiction*. New York: Penguin Books, 2012. Print.
Anderson, Martin, and Annelise Anderson. *Reagan's Secret War: The Untold Story of His Fight to Save the World from Nuclear Disaster*. New York: Crown Publishers, 2009. Print.
Anderson, Poul. *The Enemy Stars*. London: Lippincott, 1958. Print.
Anderson, Poul. "Time Patrol." *The Time Patrol*. New York: Tom Doherty Associates, 1991. Print.
Anderson, Poul. "Tomorrow's Children." *Now Begins Tomorrow*. Ed. Damon Knight. New York: Lancer, 1969.
Andrews, Robin. "Germany's Fusion Reactor Creates Hydrogen Plasma In World First." *IFL Science!* 3 Feb. 2016. Web. 4 Apr. 2016. http://www.iflscience.com/physics/germanys-fusion-reactor-creates-hydrogen-plasma-world-first

Annis, James. "An Astrophysical Explanation for the Great Silence." *Arxiv.org*. Cornell University Library. 22 Jan. 1999. Web/PDF. 18 Sep. 2016. https://arxiv.org/abs/astro-ph/9901322
"Area Defense Anti-Munitions (ADAM)." Lockheedmartin.com. Web. 1 Aug. 2016. http://www.lockheedmartin.com/us/products/ADAM.html
Armstrong, Alex. "Artificial Intelligence—Strong and Weak." *I Programmer*. 2016. Web. 10 Oct. 2016. http://www.i-programmer.info/babbages-bag/297-artificial-intelligence.html
Ashton, Kevin. "That 'Internet of Things' Thing." *RFID Journal*. 22 Jun. 2009. Web. 4 Nov. 2016. http://www.rfidjournal.com/articles/view?4986
Asimov, Isaac. *Asimov on Science Fiction*. New York: Doubleday, 1981. Print.
Asimov, Isaac. *The Bicentennial Man and Other Stories*. New York: Doubleday, 1976. Print.
Asimov, Isaac. *The Complete Robot*. New York: Doubleday, 1982. Print.
Asimov, Isaac. *The Foundation Trilogy*. New York: Doubleday, 1982. Print.
Asimov, Isaac. "Guest Commentary: The Three Laws." *Compute!* 18 (1981): 18. Print.
Asimov, Isaac. *I, Robot*. New York: Doubleday, 1950. Print.
Asimov, Isaac. *In Memory Yet Green: The Autobiography of Isaac Asimov 1920–1954*. New York: Doubleday, 1979. Print.
Asimov, Isaac. "Introduction." *The Complete Robot*. New York: Doubleday, 1982. Print.
Asimov, Isaac. "Introduction." *The Rest of the Robots*. New York: Doubleday, 1964. Print.
Asimov, Isaac. "Introduction: Robots, Comput-

ers, and Fear." *War with the Robots.* Eds. Patricia S. Warrick and Martin Greenberg. New York: Wings Books, 1983. Print.
Asimov, Isaac. "Introduction: Sailing the Wind." *Project Solar Sail.* Ed. Arthur C. Clarke. New York: Roc, 1990. Print.
Asimov, Isaac. "The Machine and the Robot." *Science Fiction: Contemporary Mythology* Eds. Patricia Warrick, Martin Harry, Joseph Olander Greenberg. New York: Harper & Row, 1978. Print.
Asimov, Isaac. "My Robots." *Robot Visions.* New York: Roc, 1991. Print.
Asimov, Isaac. "Opposite!" *The Relativity of Wrong.* New York: Pinnacle Books/Windsor Publishing Corp., 1990. Print.
Asimov, Isaac. "Reason." *The Complete Robot.* New York: Doubleday, 1982. Print.
Asimov, Isaac. *The Rest of the Robots.* New York: Doubleday, 1964. Print.
Asimov, Isaac. *Robot Visions.* New York: RoC, 1991. Print.
Asimov, Isaac. *Robots and Empire.* New York: Doubleday, 1985. Print.
Asimov, Isaac. *The Robots of Dawn.* New York: Doubleday, 1983. Print.
Asimov, Isaac, and Karen A. Frenkel. *Robots: Machines in Man's Image.* New York: Harmony Books, 1985. Print.
Asimov, Isaac, Patricia S. Warrick, and Martin H. Greenberg, eds. *War with the Robots: 28 of the Best Short Stories by the Greatest Names in 20th Century Science Fiction.* New York: Wings Books, 1991. Print.
Asimov, Isaac, and Frank White. *The March of the Millennia: A Key to Looking at History.* New York: Walker, 1991. Print.
"Author Biographies: George Griffith." *Collector's Guide Publishing Inc.* 2004. Web. http://www.apogeebooks.com/Author_Bios/geo rge_griffith.html
"BAD." *The MIT Jargon*, version 299 with appendices. Web. 14 Jan. 2016. http://magiccookie.co.uk/jargon/mit_jargon.htm
Baldick, Chris. *The Concise Oxford Dictionary of Literary Terms.* New York: Oxford University Press, 1990. Print.
Bancroft-Hinchey, Timothy. "India: Dolphins Declared Non-human Persons." *Pravda.ru.* 8 May 2013. Web. http://www.pravdareport.com/science/earth/05-08-2013/125310-dolphins_india-0/?mode=print
Banerjee, Anindita. *We Modern People: Science Fiction and the Making of Russian Modernity.* Middletown, CT: Wesleyan University Press, 2012. Print.
Banks, Iain M. *The Hydrogen Sonata.* New York: Orbit, 2012. Amazon Kindle.
Banks, Iain M. *Look to Windward.* New York: Pocket Books, 2001. Amazon Kindle.
Banks, Iain M. *Matter.* New York: Orbit, 2009. Amazon Kindle.
Banks, Iain M. *Surface Detail.* New York: Orbit, 2010. Amazon Kindle.
Baral, Susmita. "Why Aliens Shouldn't Contact Us: Stephen Hawking Warns of Intelligent Extraterrestrial Life." *International Business Times.* IBT Media. 22 Sep. 2016. Web. 25 Sep. 2016. http://www.ibtimes.com/why-aliens-shouldnt-contact-us-stephen-hawking-warns-intelligent-extraterrestrial-2420526
Barnes, Julian E. "A First Look at America's Supergun." *The Wall Street Journal.* Dow Jones and Company. 30 May 2016. Web. 9 Sep. 2016. http://www.wsj.com/articles/a-first-look-at-americas-supergun-1464359194
Barr, Marleen S. *Lost in Space: Probing Feminist Science Fiction and Beyond.* London: University of North Carolina Press, 1993. Print.
Barrat, James. *Our Final Invention: Artificial Intelligence and the End of the Human Era.* New York: Thomas Dunne Books/St. Martin's Griffin, 2013. Print.
"Battlefield Laser—Engineer Garin's Hyperboloid Has Been Built." *The Day X.* 11 Jan. 2013. Web. 23 Jul. 2016. http://the-day-x.ru/boevoj-lazer-giperboloid-inzhenera-garina-postroen.html (Russian-language source)
Bear, Greg. *Blood Music.* New York: Arbor House, 1985. Print.
Bear, Greg. *Blood Music.* Cleveland: Open Road Media Sci-Fi & Fantasy, 2014. Amazon Kindle.
Benforado, Adam. *Unfair: The New Science of Criminal Injustice.* New York: Broadway Books, 2015. Amazon Kindle.
Benford, Gregory, and David Brin. *Heart of the Comet.* New York: Bantam Books, 1986. Print.
Bennett, Jay. "Miniature 'Chipsats' Could be the First Step to Mankind Reaching Another Star." *Popular Mechanics* 2 Jun. 2016. Web. 17 Jun. 2016. http://www.popularmechanics.com/space/a21135/miniature-chipsats-could-be-the-first-step-to-reaching-the-stars/?click=my6sense
Bennett, Jay. "Pair of Planets That Could Harbor Alien Life Now Seem More Habitable Than Ever." *Popular Mechanics.* 20 Jul. 2016. Web. 23 Jul. 2016. http://www.popularmechanics.com/space/deep-space/a21947/two-planets-rocky-could-support-life/
Bennett, Jay. "The Future of the Navy's Electromagnetic Railgun Could Be a Big Step Backwards." *Popular Mechanics.* 6 Jun. 2016. Web. 10 Sep. 2016. http://www.popularmechanics.com/military/weapons/a21174/navy-electromagnetic-railgun/

Bergman, Gregory, and Josh Lambert. *Geektionary*. Avon, MA: Adams Media, 2011. Print.
Bernal, J.D. *The World, The Flesh and the Devil*. Scottsdale, AZ: Prism Key Press, 2012. Amazon Kindle.
Bester, Alfred. *Tiger! Tiger!* New York: Penguin, 1967. Print.
Bierce, Ambrose Gwinnett. "Moxon's Master." *sff.net*. Web. 30 Mar. 2016. http://www.sff.net/people/doylemacdonald/1_moxon.htm
Bignami, Giovanni, and Andrea Sommariva. *The Future of Human Space Exploration*. London: Macmillan, 2016. Print.
Blish, James. *Jack of Eagles*. London: Arrow Books, 1975. Print.
Blish, James. "Solar Plexus." *Beyond Human Ken*. Ed. Judith Merril. New York: Pennant Books, 1954. Print.
Borrell, Brendan. "Electromagnetic Railgun Blasts Off" *MIT Technology Review*. 6 Feb. 2008. Web. 8 Sep. 2016. https://www.technologyreview.com/s/409497/electromagnetic-railgun-blasts-off/
Bostrom, Nick. *Superintelligence: Paths, Dangers, Strategies*. Oxford: Oxford University Press, 2014. Print.
Boulle, Pierre. *Planet of the Apes*. New York: Gramercy Books, 1963. Print.
Boyle, Alan. "How Does Ion Propulsion Work?" *MSNBC News*. NBC Universal. Web. 5 Apr. 2009. www.msnbc.msn.com/NEWS/207293.asp
Bradbury, Ray. "Payment in Full." *Man Against Tomorrow*. Ed. William F. Nolan. New York: Avon, 1965. Print.
Bradbury, Ray. *The Toynbee Convector: Stories by Ray Bradbury*. New York: Alfred A. Knopf, 1988. Print.
Brennan, Richard P. *Dictionary of Scientific Literacy*. New York: John Wiley & Sons, 1992. Print.
Brigden, Susan. *New Worlds, Lost Worlds: The Rule of the Tudors, 1485–1603*. New York: Viking, 2000. Print.
Briggs, Katharine M. *The Anatomy of Puck*. New York: Routledge and Kegan Paul, 1959. Print.
Brin, David. "Afterword." *Heaven's Reach*. Norwalk, CT: Easton Press, 1998. Print.
Brin, David. "Are Animals Intelligent ... Enough?" *Contrary Brin*. 22 Aug. 2012. Web. 7 Jun. 2016. http://davidbrin.blogspot.com/2012/08/are-animals-intelligent-enough.html
Brin, David. *Existence*. New York: Tor Books, 2012. Print.
Brin, David. *Foundation's Triumph*. New York: HarperPrism, 1999. Print.
Brin, David. *Heaven's Reach*. Norwalk, CT: Easton Press, 1998. Print.
Brin, David. *Insistence of Vision*. Stanford, CT: Story Plant, 2016. Print
Brin, David. "It's ALIVE! And It's in Outer Spaaaace!" *Contrary Brin*. 8 Jul. 2016. Web. 8 Jul. 2016. http://davidbrin.blogspot.com/2015/07/its-alive-and-its-in-outer-spaaaace.html
Brin, David. "Science Fiction Cinema." *Contrary Brin*. 3 Aug. 2015. Web. 17 Jun. 2016. http://davidbrin.blogspot.com/2015/08/science-fiction-cinema.html
Brin, David. "The Search for Extraterrestrial Intelligence (SETI) and Whether to Send 'Messages' (METI): A Case for Conversation, Patience and Due Diligence." *Journal of the British Interplanetary Society* 67 (2014). Web. 12 Sep. 2016. http://www.davidbrin.com/meti.html
Brin, David. "Space-Launch Mass Drivers and von Neumann Machines: Science Meets Science Fiction." *Contrary Brin*. 21 Mar. 2012. Web. 10 Sep. 2016. http://davidbrin.blogspot.fr/2012/03/space-launch-mass-drivers-and-von.html
Brin, David. *Startide Rising*. Norwalk, CT: Easton Press, 1993. Print.
Brin, David. *Through Stranger Eyes: Reviews, Introductions, Tributes & Iconoclastic Essays*. Ann Arbor, MI: Nimble Books, 2008. Print.
Brooks, Rodney A. *Flesh and Machines: How Robots Will Change Us*. New York: Pantheon Books, 2002. Print.
Brown, Fredric. "All Good Bems." *The Second Fredric Brown Megapack: 27 Classic Science Fiction Stories*. Rockville, MD: Wildside Press, 2014. Amazon Kindle.
Brown, Fredric. "The Star Mouse." *The Fredric Brown Megapack: 33 Classic Science Fiction Stories*. Rockville, MD: Wildside Press, 2013. Amazon Kindle.
Bujold, Lois McMaster. "Afterword." *Miles, Mystery and Mayhem*. New York: Baen, 2003. Print.
Bujold, Lois McMaster. *Brothers in Arms*. Riverdale, NY: Baen, 1989. Print.
Bujold, Lois McMaster. *Cetaganda*. New York: Baen, 1996. Print.
Bujold, Lois McMaster. *Ethan of Athos*. New York: Baen, 1986. Print.
Bujold, Lois McMaster. "The Mountains of Mourning." In *Young Miles*. New York: Baen, 1997. Print.
Bujold, Lois McMaster. *The Vor Game*. In *Young Miles*. New York: Baen, 1997. Print.
Bujold, Lois McMaster. *The Warrior's Apprentice*. In *Young Miles*. New York: Baen, 1997. Print.
Byron, George Gordon Noel. *Manfred*. In *Byron's*

Works Cited

Poetry. Ed. Frank D. McConnell. New York: W.W. Norton, 1978. Print.

Campbell, John W. *Islands of Space*. New York: Ace Books, 1931. Project Gutenberg e-book.

Campbell, Joseph. *The Hero with a Thousand Faces*. New York: Meridian Books, 1971. Print.

Čapek, Karel. *R.U.R.* in *Čapek: Four Plays*. London: Methuen Drama, 1999. Print.

Carpineti, Alfredo. "Ion Engine Breakthrough Could Take Us to Mars at a Fraction of the Fuel." *IFL Science!* 28 Oct. 2015. Web. 15 Mar. 2016. http://www.iflscience.com/ion-engine-breakthrough-could-take-us-mars-fraction-fuel

Carter, Stephen L. "A Computer Wins at Go, and This Human Is Disappointed." *BloombergView*. Bloomberg. 11 Mar. 2016. Web. 24 May 2016. https://www.bloomberg.com/view/articles/2016-03-11/google-s-alphago-wins-and-this-human-is-disappointed

Cavas, Christopher P. "Destroyer Zumwalt Breaks Down and Gets Tow in Panama Canal." *Navy Times*. A Military Times and Sightline Media Group. 22 Nov. 2016. Web. 19 Jan. 2017. https://www.navytimes.com/articles/destroyer-zumwalt-breaks-down-and-gets-tow-in-panama-canal

Cellan-Jones, Rory. "Stephen Hawking Warns Artificial Intelligence Could End Mankind." *BBC*. BBC News Services. 2 Dec. 2014. Web. 5 Aug. 2016. http://www.bbc.com/news/technology-30290540

Chang, Kenneth. "NASA Announces Extension of 9 Spacecraft Missions." *The New York Times*. 2 Jul. 2016. Web. 8 Jul. 2016. http://www.nytimes.com/2016/07/03/science/nasa-extends-spacecraft-missions.html?action=click&contentCollection=Science&module=RelatedCoverage&region=EndOfArticle&pgtype=article&_r=0

Chang, Kenneth. "NASA's Juno Spacecraft Enters Into Orbit Around Jupiter." *The New York Times*. 5 Jul. 2016. Web. 6 Jul. 2016. http://www.nytimes.com/2016/07/05/science/juno-enters-jupiters-orbit-capping-5-year-voyage.html?_r=0

Chang, Lulu. "Finders Keepers Is the New Rule When It Comes to Asteroid Mining." *Digital Trends*. 26 Nov. 2015. Web. 2 Aug. 2016. http://www.digitaltrends.com/cool-tech/commercial-space-launch-competetiveness-act-2015-asteroid-mining/

Chapman, James, and Nicholas J. Cull. *Projecting Tomorrow: Science Fiction and Popular Cinema*. London: I.B. Tauris, 2013. Print.

Choueiri, Edgar Y. "A Critical History of Electric Propulsion: The First 50 Years (1906–1956)." *Journal of Propulsion and Power* 20.2 (2004): 193–203. Print.

Choueiri, Edgar Y. "New Dawn for Electric Rockets." *Scientific American* Feb. 2009: 58-65. Print.

Clarke, Arthur. C. *Childhood's End*. New York: Harcourt, Brace & World, 1953. Print.

Clarke, Arthur C. *Earthlight*. New York: Ballantine Books, 1983. Print.

Clarke, Arthur C. *Profiles of the Future*. New York: Bantam Books, 1971. Print.

Clarke, Arthur C. *Profiles of the Future*. New York: Holt, Rinehart and Winston, 1984. Print.

Clarke, Arthur C. ed. *Project Solar Sail*. New York: Roc, 1990. Print.

Clarke, Arthur C. *The Songs of Distant Earth*. New York: Del Rey, 1986. Print.

Clarke, Arthur C. "The Star." *The Collected Stories of Arthur C. Clarke*. New York: Tor, 2001. Print.

Clarke, Arthur C. "The Wind from the Sun." *Project Solar Sail*. Ed. Arthur C. Clarke. New York: Roc, 1990. Print.

Clarke, Frederick S., and Steve Rubin. "Making Forbidden Planet." *Cinefantastique* 8.2 (1979): 4–66. Print.

Clarke, Steve. "BBC Conjured up More Merlin." *Variety*. Penske Business Media. 15 Oct. 2010. Web. 19 Aug. 2016. http://variety.com/2010/biz/news/bbc-conjures-up-more-merlin-1118026310/

Clery, Daniel. "ITER Fusion Project to Take at Least 6 Years Longer than Planned." *Science*. American Association for the Advancement of Science. 19 Nov. 2015. Web. 5 Mar. 2016. http://www.sciencemag.org/news/2015/11/iter-fusion-project-take-least-6-years-longer-planned

Clute, John, and John Grant. *The Encyclopedia of Fantasy*. New York: St. Martin's Griffin, 1997. Print.

Clute, John, David Langford, Peter Nicholls, and Graham Sleight, eds. *The Encyclopedia of Science Fiction* (online edition). SF-Encyclopedia. SFE, Ltd. 2011. Web. 6 Nov. 2016. http://www.sf-encyclopedia.com/

Clute, John, and Peter Nicholls. *The Encyclopedia of Science Fiction*. New York: St. Martin's Press, 1993. Print.

Collen, Alanna. *10% Human: How Your Body's Microbes Hold the Key to Health and Happiness*. New York: Harper, 2015. Print.

Collins English Dictionary Complete & Unabridged. Digital edition. New York: HarperCollins Publishers, 2012. Web. http://www.dictionary.com/browse/speculative-fiction

"Computer Wins Final Game Against S. Korean

Go Master." *Yahoo!Tech*. Yahoo! 15 Mar. 2016. Web. 25 May 2016. https://www.yahoo.com/tech/computer-wins-final-game-against-korean-master-095605123.html

Cook, Michael. "Stop Fretting about 3-Parent Embryos and Get Ready for 'Multiplex Parenting.'" *Bioedge*. 9 Mar. 2014. Web. 15 Jul. 2016. http://www.bioedge.org/bioethics/bioethics_article/10882

Coren, Stanley. *The Intelligence of Dogs: A Guide to the Thoughts, Emotions, and Inner Lives of Our Canine Companions*. New York: Bantam, 1994. Print.

Cornish, Edward. *Futuring: The Exploration of the Future*. Bethesda, MD: World Future Society, 2004. Print.

Coulter, Dauna. "A Brief History of Solar Sails." *NASA Science*. NASA. 2008. Web. 20 Jun. 2016. http://science.nasa.gov/science-news/science-at-nasa/2008/31jul_solarsails/

Cousteau, Jacques. *The Ocean World*. New York: Abradale Press/Harry N. Abrams, 1979. Print.

Csicsery-Ronay, Istvan. "The SF of Theory: Baudrillard and Haraway." *Science Fiction Studies*. 18.55 (1991). www.depauw.edu/sfs/backissues/55/icr55art.htm.

Curry, Colleen. "Scientists Announce Successful Teleportation, But Can You Beam Yourself to the Moon?" *abcNews*. 30 May 2014. Web. http://abcnews.go.com/Technology/scientists-announce-successful-teleportation-beam-moon/print?id=23931524

Cutas, Daniela, and Anna Smajdor. "'I Am Your Mother and Your Father!' In Vitro Derived Gametes and the Ethics of Solo Reproduction." *Health Care Analysis* 11 Mar. 2016. Web. 2 Aug. 2016. http://link.springer.com/article/10.1007/s10728-016-0321-7

Darling, David. *Teleportation: The Impossible Leap*. Hoboken, NJ: John Wiley and Sons, 2005. Print.

Darrow, Clarence, and Will Durant. *Debate: Is Man a Machine?* New York: League for Public Discussion, 1927. Print.

Davies, Paul. *About Time: Einstein's Unfinished Revolution*. New York: Simon & Schuster, 1995. Print.

Davies, Paul. *The Eerie Silence: Renewing Our Search for Alien Intelligence*. New York: Mariner Books, 2010. Print.

Davis, Jason. "Sailing to the World's Most Famous Comet." *The Planetary Society*. Web. 20 Jun. 2016. http://sail.planetary.org/story-part-1.html

Davis, Stan. *Lessons from the Future: Making Sense of a Blurred World*. Oxford: Capstone, 2001. Print.

de Bergerac, Cyrano. *Other Worlds: The Comical History of the States and Empires of the Moon and the Sun*. Oxford: Oxford University Press, 1965. Print.

de Camp, L. Sprague. *Lest Darkness Fall*. New York: Ballantine Books, 1949. Print.

DeGroot, Gerard J. *The Bomb: A Life*. Cambridge, MA: Harvard University Press, 2005. Print.

Dekker, Thomas, John Ford, and William Rowley. *The Witch of Edmonton. Jacobean and Caroline Comedies*. Ed. R. G. Lawrence. London: J. M. Dent, 1973. Print.

de Lint, Charles. "Timeskip." *The Very Best of Charles de Lint*. Triskell Press, 2014. Amazon Kindle.

Del Rey, Lester. "The Day Is Done." *Where Do We Go from Here: 17 Great Science Fiction Classics*. Ed. Isaac Asimov. Greenwich, CT: Fawcett Publications, Inc., 1971. Print.

Descartes, Rene. *Meditations on First Philosophy*. Cambridge: Cambridge University Press, 1911. Web. 6 Jul. 2016. http://www.sacred-texts.com/phi/desc/med.txt

De Vries, Hugo. *Species and Varieties: Their Origin by Mutation*. Chicago: Open Court, 1905. Print.

Diamandis, Peter H., and Steven Kotler. *Abundance: The Future Is Better Than You Think*. New York: Free Press, 2014. Print.

Dick, Philip K. *The Collected Stories of Philip K. Dick*, vol. 3. New York: Carol Publishing Group, 1991. Print.

Dick, Philip K. "The Golden Man." *The Collected Stories of Philip K. Dick*, vol. 3. New York: Carol Publishing Group, 1991. Print.

Dickson, Gordon R. *Wolfling*. New York: Baen, 1969. Print.

Dowling, David. *Fictions of Nuclear Disaster*. Iowa City: University of Iowa Press, 1987. Print.

Dozois, Gardner, and Jack Dann, eds. *Futures Past*. New York: Baen, 2013. Baen e-book.

Drexler, K. Eric. *Engines of Creation: The Coming Era of Nanotechnology*. Harpswell, ME: Anchor Publishing, 1997. Print.

Drexler, K. Eric. "Introduction to Nanotechnology." *Prospects in Nanotechnology: Toward Molecular Manufacturing*. Eds. Markus Krummenacker and James Lewis. New York: John Wiley & Sons, Inc., 1995. Print.

Dvorsky, George. "The Ethics of Animal Enhancement." *Sentient Developments*. 26 Jul. 2011. Web. http://www.sentientdevelopments.com/2011/07/ethics-of-animal-enhancement.html

Dvorsky, George. "How to Build an Artificial Womb." *io9*. 19 Apr. 2013. Web. 10 Jun. 2016. http://io9.gizmodo.com/how-to-build-an-artificial-womb-476464703

Dvorsky, George. "Why Asimov's Three Laws of Robotics Can't Protect Us." *io9*. 28 Mar. 2014. Web. http://io9.com/why-asimovs-three-laws-of-robotics-cant-protect-us-1553665410

Dworschak, Manfred. "The Chernobyl Conundrum: Is Radiation As Bad As We Thought?" *Spiegel Online*. Der Spiegel. 26 Apr. 2016. Web. 28 Apr. 2016. http://www.spiegel.de/international/world/chernobyl-hints-radiation-may-be-less-dangerous-than-thought-a-1088744.html

Dyson, George. *Project Orion: The True Story of the Atomic Spaceship*. New York: Owl Books, 2003. Print.

Edmonds, David, and John Eidinow. *Bobby Fischer Goes to War*. New York: HarperCollins Publishers, 2004. Print.

Egan, Greg. *Diaspora*. New York: HarperPrism, 1998. Print.

Eiseley, Loren. *The Immense Journey*. New York: Vintage Books, 1959. Print.

Eiseley, Loren. *The Invisible Pyramid: A Humanist Account of the Space Age*. London: Rupert Hart-Davis, 1970. Print.

Engelbert, Phillis, and Diane L. Dupuis. *The Handy Space Answer Book*. Detroit: Visible Ink, 1998. Print.

Enriquez, Juan, perf. *TED Talks Science and Wonder*. Dir. Linda Mendoza. ITVS, 2016. Television.

Enriquez, Juan, and Steve Gullans. *Evolving Ourselves: How Unnatural Selection and Nonrandom Mutation Are Changing Life on Earth*. New York: Current, 2015. Print.

"Estate of William F. Jenkins v. Paramount Pictures Corp." Civil Action Lawsuit. March 20 March 2000. Web. http://en.wikisource.org/wiki/Estate_of_William_F._Jenkins_v._Paramount_Pictures_Corp

Ettinger, Robert C. W. *The Prospect of Immortality*. Clinton Township, MI: Cryonics Institute. PDF.

Eveleth, Rose. "Robots: Is the Uncanny Valley Real?" BBC.com. 2 Sep. 2013. Web. http://www.bbc.com/future/story/20130901-is-the-uncanny-valley-real

Fan, Shelley. "Scientists Connect Brain to a Basic Tablet—Paralyzed Patient Googles with Ease." *SingularityHUB*. Singularity University. 25 Oct. 2015. Web. 13 May 2016. http://singularityhub.com/2015/10/25/scientists-connect-brain-to-a-basic-tablet-paralyzed-patient-googles-with-ease/

Farrand, Phil. *The Nitpicker's Guide for Next Generation Trekkers*, vol. 2. New York: Del, 1995. Print.

Fetzer, Leland. "Introduction." *Pre-Revolutionary Russian Science Fiction: An Anthology (Seven Utopias and a Dream)*. Ed. Leland Fetzer. Ann Arbor, MI: Ardis, 1982. Print.

Feynman, Richard P. "There's Plenty of Room at the Bottom." Ed. Jeffrey Robbins. *The Pleasure of Finding Things Out: The Best Short Works of Richard P. Feynman*. Cambridge, MA: Perseus Books, 1999. Print.

Ford, Paul. "The Last Museum." *Motherboard*. 8 June 2015. Web. 17 Aug. 2016. http://motherboard.vice.com/read/last-museum

Ford, Paul. "Our Fear of Artificial Intelligence." *MIT Technology Review*. 11 Feb. 2015. Web. 6 Oct. 2016. https://www.technologyreview.com/s/534871/our-fear-of-artificial-intelligence/

Forde, Pat. "In Spirit." *Analog: Science Fiction and Fact*. Dell Magazines. Sep. 2002. Web. 2 Aug. 2016. https://www.analogsf.com/neb ulas03/spirit.shtml

Frank, Pat. *Mr. Adam*. New York: Harper Perennial, 2016. Print.

Freitas, Robert A. " The Legal Rights of Extraterrestrials." rfreitas.com. 1977/2008. Web. 29 Jul. 2016. http://www.rfreitas.com/Astro/LegalRightsOfETs.htm

Freud, Sigmund. "The Uncanny." Trans. Alix Strachey. mit.edu. Web. http://web.mit.edu/allanmc/.com/freud1.pdf

Friedman, Louis. *Starsailing: Solar Sails and Interstellar Travel*. New York: John Wiley and Sons, 1988. Print.

"Fusion Reactor Designed in Hell Makes Its Debut." *Youtube*. Science Magazine. 22 Oct. 2015. Web. 7 March 2016. https://www.youtube.com/watch?v=u-fbBRAxJNk&lc=z12at1gaqxzruzhhm04cjv04xremhzopykk0k

Gabaldon, Diana. "Introduction." *Outlander: 20th Anniversary Collector's Edition*. New York: Delacorte Press, 2011. Print.

Garreau, Joel. *Radical Evolution: The Promise and Peril of Enhancing Our Minds, Our Bodies—and What It Means to Be Human*. New York: Doubleday, 2005. Print.

Garucho, Roberto. "Navy Stealth Ship USS Zumwalt Destroyer Sails to San Diego." *The Cubic Lane*. 10 Sep. 2016. Web. 11 Sep. 2016. http://cubiclane.com/2016/09/10/navy-stealth-ship-uss-zumwalt-destroyer-sails-to-san-diego/

George, Peter. *Red Alert*. New York: Rosetta Books, 1958. Amazon Kindle.

"Germans Have Successfully Tried out the Hyperboloid of Engineer Garin—a High-energy Laser—Our Scientists Remain Quiet." *Neftegas.Ru*. 16 Sep. 2012. Web. 1 Aug. 2016. http://neftegaz.ru/news/view/104231-Nemtsy-uspeshno-ispytali-giperboloid-inzhenera-

Garina-vysokoenergeticheskiy-lazer.-Nashi-molchat (Russian language source)

Ghamari-Tabrizi, Sharon. *The Worlds of Herman Kahn: The Intuitive Science of Thermonuclear War*. Cambridge, MS: Harvard University Press, 2005. Print.

Gibbs, Samuel. "Musk, Wozniak and Hawking Urge Ban on Warfare AI and Autonomous Weapons." *The Guardian*. 27 Jul. 2015. Web. 5 Oct. 2016. https://www.theguardian.com/technology/2015/jul/27/musk-wozniak-hawking-ban-ai-autonomous-weapons#t

Gibson, William. *Neuromancer*. New York: Ace Books, 1986. Print.

Gilster, Paul. "Science, Fiction and the Sail." *Centauri Dreams: Imagining and Planning Interstellar Exploration*. 8 Jun. 2012. Web. 17 Jun. 2016. http://www.centauri-dreams.org/?p=23167

Girin, Alexander. "Formation and Development of Military Uniform Clothing for the Service Personnel of the Armed Forces of Russian Federation." *Ruguard*. 3 Feb. 2012. Web. http://ruguard.ru/forum/index.php/topic,670.0.html (Russian language source)

Gleick, James. *Genius: The Life and Science of Richard Feynman*. New York: Pantheon Books, 1992. Print.

"Glenn Contributions to Deep Space 1." *NASA*. 21 May 2008. Web. 10 Jun. 2016. http://www.nasa.gov/centers/glenn/about/history/ds1.html

Golden, Christie. *The Farther Shore*. New York: Pocket Books, 2003. Print.

Goldenfeld, Nigel. "Indivi-duality." *This Idea Must Die: Scientific Theories That Are Blocking Progress*. Ed. John Brockman. New York: Harper Perennial, 2015. Print.

Granholm, Jackson W. "How to Design a Kludge." *Datamation* February 1962. Web. http://neil.franklin.ch/Jokes_and_Fun/Kludge.html

Grässler, Bernd, David Levitz, and Ben Knight. "Tens of Thousands of Germans Take to the Streets in Anti-nuclear Protest." *DW*. Deutsche Welle. 13 March 2011. Web. 8 Feb. 2016. http://www.dw.com/en/tens-of-thousands-of-germans-take-to-the-streets-in-anti-nuclear-protest/a-14907859

Graves, Robert. *The Greek Myths*. London: Moyer Bell, 1988. Print.

Grazier, Kevin R. and Stephen Cass. *Hollyweird Science: From Quantum Quirks to the Multiverse*. New York: Springer, 2015. Print.

Great Soviet Encyclopedia. Ed. A. M. Prokhorov. New York: Macmillan, 1980. Print.

Green, Jonathon. *Green's Dictionary of Slang*, vol. 2. London: Chambers, 2010. Print.

Greene, Robert. *Friar Bacon and Friar Bungay*. Ed. Daniel Seltzer. London: Edward Arnold, 1964. Print.

Griffith, George C. *The World Masters*. London: John Long, 1903. Project Gutenberg e-book.

Griffith, Mary. *Three Hundred Years Hence*. Seattle: Amazon Digital Services, 2015. Amazon Kindle.

Grinspoon, David. "Who Speaks for Earth?" *Seed*. Seed Media Group. 15 Jul. 2015. Web. 28 Sep. 2016. http://seedmagazine.com/content/article/who_speaks_for_earth/

Grush, Loren. "At Last, Bill Nye's LightSail Deploys Its Solar Sails In Space." *Popular Science*. Bonnier Corporation. 8 Jun. 2015. Web. 5 Jun. 2016. http://www.popsci.com/last-light-sail-deploys-its-solar-sails-space

Gunn, James. "Introduction." Charles Pellegrino and George Zebrowski. *The Killing Star*. Norwalk, CT: Easton Press, 1995. Print.

Gunn, James. "Introduction." Charles Sheffield. *Tomorrow and Tomorrow*. Norwalk, CT: Easton Press, 1997. Print.

Gunn, James. *The Listeners*. New York: Open Road Media, 2014. Amazon Kindle.

Haley, Andrew G. *Space Law and Government*. New York: Appleton-Century-Crofts, 1963. Print.

Hall, J. Storrs. *Nanofuture: What's Next for Nanotechnology*. New York: Prometheus, 2005. Print.

Hall, J. Storrs. "On Certain Aspects of Utility Fog." *Pivot.net*. 11 Sep. 2005. Web. 4 Nov. 2016. http://www.pivot.net/~jpierce/aspects_of_ufog.htm

Hall, J. Storrs. "Utility Fog: The Stuff that Dreams Are Made Of." *Kurzweil: Accelerating Intelligence*. 5 Jul. 2001. Web. 4 Nov. 2016. http://www.kurzweilai.net/utility-fog-the-stuff-that-dreams-are-made-of

Hamilton, Edmond. *Captain Future: Man of Tomorrow*. California: Pulpville Press, 2005. Print.

Harris, Anthony. *Night's Black Agents: Witchcraft and Magic in Seventeenth-Century English Drama*. Totowa, NJ: Rowman and Littlefield, 1980. Print.

Harris, Brent A. "Foreword." *Tales from Alternate Earths*. Inklings Press, 2016. Amazon Kindle.

Harris, John. "Misleading Talk of 'Three-parent Babies' Helps no One." *The Guardian*. 19 Sept. 2012. Web. 1 Aug. 2016. https://www.theguardian.com/commentisfree/2012/sep/19/misleading-three-parent-babies-gene-therapy

Healey, Raymond J., and J. Francis McComas. *Adventures in Time and Space: An Anthology of Science Fiction Stories*. New York: Random House, 1957. Print.

Hearne, Thomas. *Ductor Historicus: Or, a Short System of Universal History, and an Introduction to the Study of It*. 3rd ed. London: H. Clark, 1714. Print.
Heinlein, Robert A. "All You Zombies—." Seattle: Amazon Digital Services, 2012. Amazon Kindle.
Heinlein, Robert A. *Citizen of the Galaxy*. New York: Spectrum Literary Agency, 2013. Amazon Kindle.
Heinlein, Robert. *The Door Into Summer. A Heinlein Trio*. New York: Nelson Doubleday, 1957. Print.
Heinlein, Robert A. *Grumbles from the Grave*. New York: Ballantine, 1989. Print.
Heinlein, Robert A. *Have Space Suit—Will Travel*. New York: Ballantine Books, 1958. Print.
Heinlein, Robert A. *Methuselah's Children*. Riverdale, NY: Baen, 1958. Print.
Heinlein, Robert A. *The Moon Is a Harsh Mistress*. New York: Tom Doherty Associates, 1966. Print.
Henig, Robin Marantz. "Death by Robot." *The New York Times Magazine* 9 Jan. 2015. Web. http://www.nytimes.com/2015/01/11/magazine/death-by-robot.html
Hoffmann, E.T.A. *The Devil's Elixirs*, vol. 1 and 2. Edinburgh: William Blackwood, 1829. Project Gutenberg e-book.
Hoffmann, E.T.A. "Mademoiselle de Scudery." *The Tales of Hoffman*. Norwalk, CT: Easton Press, 1992. Print.
Hoffmann, E.T.A. "Nutcracker and the King of Mice." *The Serapion Brethren*, vol. 1. London: George Bell and Sons, 1908. Project Gutenberg e-book.
Holmes, Oliver Wendell. *Elsie Venner*. New York: Grosset & Dunlap, 1861. Print.
Holt, Nathalia. *Rise of the Rocket Girls: The Women Who Propelled Us, from Missiles to the Moon to Mars*. New York: Little, Brown, 2016. Print.
Homer. *The Iliad*. Samuel Butler, transl. Roslin, NY: Walter J. Black, 1944. Print.
Homer. *The Odyssey*. Samuel Butler, transl. Roslin, NY: Walter J. Black, 1944. Print.
Homer. *The Odyssey*. W. H.D. Rouse, transl. New York: Signet, 1937. Print.
Homer. *The Odyssey*. Walter Shewring, transl. Oxford: Oxford University Press, 1980. Print.
Hong, Guosong, Tian-Ming Fu, Tao Zhou, Thomas G. Schuhmann, Jinlin Huang, and Charles M. Lieber. "Syringe Injectable Electronics: Precise Targeted Delivery with Quantitative Input/Output Connectivity." *NanoLetters* 15 (2015): 6979–6984. Web. May 16, 2016. pubs.acs.org/NanoLett
"How to Make a Man." 2015. *Internet Broadway Database*. Web. 7 Oct. 2015. http://ibdb.com/production.php?id=2290
Howard, Philip N. *Pax Technica: How the Internet of Things May Set Us Free or Lock Us Up*. New Haven, CT: Yale University Press, 2015. Print.
Howell, Elizabeth. "Ikaros: First Successful Solar Sail." Space.com. 7 May 2014. Web. 20 Jun. 2016. http://www.space.com/25800-ikaros-solar-sail.html
Hsu, Jeremy. "Could the Navy Ever Build a Flying Aircraft Carrier?" *Live Science*. 4 May 2012. http://m.livescience.com/20117-navy-flying-aircraft-carrier.html
"Internet of Things Global Standards Initiative." *ITU*. 2016. Web. 4 Nov. 2016. http://www.itu.int/en/ITU-T/gsi/iot/Pages/default.aspx
"Interstellar Travel: Episode 145." *Astronomy Cast*. 28 Jul. 2009. Podcast. 5 Jun. 2016. http://www.astronomycast.com/2009/07/ep-145-interstellar-travel/
"Ion Propulsion: Over 50 Years in the Making." *NASA*. Web. 19 May 2016. http://science.nasa.gov/science-news/science-at-nasa/1999/prop06apr99_2/
Jacobsen, Annie. *The Pentagon's Brain: An Uncensored History of DARPA, America's Top Secret Military Research Agency*. New York: Little, Brown, 2015. Print.
The Jargon File. "Kludge." Catb.org. Version 4.4.7. Web. http://catb.org/jargon/html/K/kludge.html
Jenkin, John G. "Atomic Energy Is 'Moonshine': What Did Rutherford Really Mean?" *Physics in Perspective* 13 (2011): 128–145. Print.
Jentsch, Ernst. "On the Psychology of the Uncanny." 1906. art3idea. Web. http://art3idea.psu.edu/locus/Jentsch_uncanny.pdf
Johnson, Les. "Solar and Beamed Energy Sails." *Going Interstellar*. Eds. Les Johnson and Jack McDevitt. New York: Baen, 2012. Baen e-book.
Johnson, Terry. "Where Is My Uterine Replicator (AKA Artificial Womb)?" *io9*. 8 May 2008. Web. 1 Jul. 2016. http://io9.gizmodo.com/385976/where-is-my-uterine-replicator-aka-artificial-womb
Jonson, Ben. *The Alchemist. Elizabethan Plays*. Ed. Hazelton Spencer. Boston: D. C. Heath, 1933. Print.
Jules-Verne, Jean. *Jules Verne: A Biography*. New York: Taplinger, 1976. Print.
Jung, Carl G. *Man and His Symbols*. New York: Dell, 1968. Print.
Jung, Carl G. *Psychology and Education*. Princeton, NJ: Princeton University Press, 1954. Print.

"Juno Spacecraft and Instruments." *NASA*. Web. 5 Jul. 2016. https://www.nasa.gov/mission_pages/juno/spacecraft/index.html

"Juno Spacecraft to Carry Three Figurines to Jupiter Orbit" *NASA*. 3 Aug. 2011. Web. 7 Jul. 2016. http://www.nasa.gov/mission_pages/juno/news/lego20110803.html

Joyce, James. *Finnegans Wake*. Ware, Hertfordshire: Wordsworth Editions, 2012. Print.

Kahn, Herman. *On Thermonuclear War*. Princeton, NJ: Princeton University Press, 1960. Print.

Kaku, Michio. "How Jupiter Saves Earth from Destruction." *NBC-2*. Frankly Media. 5 Jul. 2016. Web. 5 Jul. 2016. http://www.nbc-2.com/story/32377386/how-jupiter-saves-earth-from-destruction

Kaufman, Scott. *Project Plowshare: The Peaceful Use of Nuclear Explosives in Cold War America*. Ithaca, NY: Cornell University Press, 2012. Print.

Kazantsev, Alexander. "Rays of Darkness." *Engineer Garin and His Death Ray*. Alexei Tolstoi. Moscow: Raduga Publishers, 1987. Print.

Keegan, Rebecca. *The Futurist: The Life and Films of James Cameron*. New York: Crown Archetype, 2009. Print.

Kiderra, Inga. "Your Brain on Androids." *UC San Diego News Center*. 14 Jul. 2011. Web. http://ucsdnews.ucsd.edu/archive/newsrel/soc/20110714BrainAndroids.asp#

Kipling, Rudyard. *The Jungle Books*. Norwalk, CT: Easton Press, 1980.

Kippenhahn, Rudolf. *Bound to the Sun: The Story of Planets, Moons, and Comets*. New York: W. H. Freeman, 1990. Print.

Knowles, Christopher. *Our Gods Wear Spandex: The Secret History of Comic Book Heroes*. San Francisco: Weiser Books, 2007. Print.

Kofsky, Michael. "NASA's Juno Probe Completes 5-year Journey to Jupiter." *USA Today*. 5 Jul. 2016. Web. 6 Jul. 2016.

Krauss, Lawrence. *The Physics of Star Trek*. New York: Basic Books, 1995. Print.

Krulik, Theodore. "A Few Words from Roger Zelazny." *Tor*. 3 Aug. 2016. Web. 7 Nov. 2016. www.tor.com.

Kurzweil, Ray. "Implantable Brain Electronics Is Here." *Kurzweil Accelerating Intelligence*. KurzweilAINetwork. 10 Jun. 2015. Web. 17 May 2016. http://www.kurzweilai.net/implantable-brain-electronics-is-here

Kurzweil, Ray. *The Singularity Is Near: When Humans Transcend Biology*. New York: Viking, 2005. Print.

Labriola, Patrick. "Edgar Allan Poe and E. T. A. Hoffmann: The Double in 'William Wilson' and *The Devil's Elixirs*." *The International Fiction Review* 29.1 and 2 (2002). Web. http://journals.hil.unb.ca/index.php/IFR/article/view/7718/8775

Lang, Andrew. "Suspended Animation." *The Collected Works of Andrew Lang*. New York: PergamonMedia, 2015. Amazon Kindle.

Lanouette, William. *Genius in the Shadows: A Biography of Leo Szilard, the Man behind the Bomb*. New York: Skyhorse Publishing, 2013. Amazon Kindle.

Lawson, Dominic. *End Game: Kasparov vs. Short*. New York: Harmony Books, 1994. Print.

Leeming, David Adams. *Mythology: The Voyage of the Hero*. New York: J. B. Lippincott, 1973. Print.

"LEGO Minifigures on NASA's Juno Jupiter Probe Inspire Design Challenge." Collectspace.com. 4 Jul. 2015. Web. 5 Jul. 2016. http://www.collectspace.com/news/news-070416a-juno-jupiter-lego-minifigures.html

Leinster, Murray. *First Contacts: The Essential Murray Leinster*. Ed. Joe Rico. Framingham, MA: NESFA Press, 1998. Print.

Leinster, Murray. "A Logic Named Joe." *War with the Robots: 28 of the Best Short Stories by the Greatest Names in 20th Century Science Fiction*. Ed. Issac Asimov, Patricia S. Warrick, and Martin H. Greenberg. New York: Wings Books, 1991. Print.

Leinster, Murray. "Sidewise in Time." *First Contacts: The Essential Murray Leinster*. Ed. Joe Rico. Framingham, MA: NESFA Press, 1998. Print.

Levin, Ira. *The Stepford Wives*. New York: Random House, 1972. Print.

Levin, Laura Victoria. "Introduction." *The Strange Case of Dr. Jekyll and Mr. Hyde*. Robert Louis Stevenson. New York: Barnes and Noble, 1995. Print.

Levy, Steven. "Big Blue's Hand of God." *Newsweek* 18 May 1997. Web. 25 May 2016. http://www.newsweek.com/big-blues-hand-god-173076

Levy, Steven and Bill Powell. "Man vs. Machine." *Newsweek* 5 May 1997. Web. 27 May 2016. http://faculty.ycp.edu/~dweiss/phl221_intro/man_vs_machine.htm

Lewis, C. S. *The Discarded Image*. Cambridge, UK: Cambridge University Press, 1964. Print.

Ley, Willy. "A Station in Space." *Across the Space Frontier*. Ed. Cornelius Ryan. New York: Viking, 1952. Print.

Lidbetter, Guy. "The Speed and Future of Technology Change: The Times They Are a Changing." *HuffPost Tech*. 7 December 2012. Web. 27 May 2015. http://www.huffingtonpost.co.uk/guy-lidbetter/the-speed-and-future-of-t_1_b_1667215.html?view=print&comm_ref=false

Linden, David J. *The Accidental Mind: How Brain Evolution Has Given Us Love, Memory, Dreams, and God*. Cambridge, MA: Belknap Press of Harvard University Press, 2007. Print.
Lipson, Hod, and Melba Kurman. *Fabricated: The New World of 3D Printing*. Indianapolis, IN: John Wiley & Sons, 2013. Print.
Livio, Mario. "Extraterrestrial Civilizations: Coming of Age in the Milky Way." *NASA News On Point Cast*. 10 Dec. 1998. Web. 1 Jan. 1998. http://127.0.0.1:15841/v1?catid=16 258051& md5=4afda0515e4c0f35a3600e8 598c8472d
"Lockheed Martin Demonstrates ADAM Ground-Based Laser System Against Military-Grade Small Boats." *Lockheedmartin.com*. May 2014. Web. 2 Aug. 2016. http://www.lockheedmartin.com/us/news/press-releases/2014/may/0507-ss-adam.html
Lowry, Rich. "Gingrich: The Republican Clinton." *National Review Online* 24 Jan. 2012. Web. http://www.nationalreview.com/articles/288989/gingrich-republican-clinton-rich-lowry
MacGregor, Neil. *Shakespeare's Restless World: A Portrait of an Era in Twenty Objects*. New York: Viking, 2012. Print.
"MacGyver." *IMDb*. Web. http://www.IMDb.com/title/tt0088559/
Madrigal, Alexis. "Autonomous Robots Invade Retail Warehouses." *Wired*. Wired. 27 Jan. 1999. Web. 1 Mar. 2016.
Majer, Peter, and Cathy Porter. "Introduction." *Karel Čapek: Four Plays*. London: Bloomsbury, 1999. Print.
Mann, George, ed. *The Mammoth Encyclopedia of Science Fiction*. New York: Carroll & Graf Publishers, 2001. Print.
Mann, Helen. "Debating the Search for Alien Life." *Day 6*. CBC/Radio Canada. 24 Jul. 2015. Radio Broadcast/Web. 15 Aug. 2016. http://www.cbc.ca/radio/day6/episode-243-talking-to-aliens-kanye-west-airbnb-vs-the-apartment-hunter-and-more-1.3159928/debating-the-search-for-alien-life-1.3161754
Marcus, Gary. *Kluge: The Haphazard Evolution of the Human Mind*. New York: Mariner Books, 2009. Amazon Kindle.
Marlowe, Christopher. *Christopher Marlowe's Doctor Faustus: Text and Major Criticism*. Ed. Irving Ribner. New York: Odyssey Press, 1966. Print.
Marston, Wendy. "Wonder Wear." *Discover Magazine* 21.1 (2000): 46–48. Print.
McCarthy, Wil. *The Collapsium*. New York: Del Rey Books, 2000. Print.
McCarthy, Wil. *Hacking Matter: Levitating Chairs, Quantum Mirages, and the Infinite Weirdness of Programmable Atoms*. New York: Basic Books, 2003. Print.
McIntosh, Will. "Bridesicle." *Asimov's Science Fiction*. Jan. 2009. Web. 15 Aug. 2016. http://will.tip.dhappy.org/blog/Compression%20Trees/.../book/by/Will%20McIntosh/Bridesicle/Will%20McIntosh%20-%20Bridesicle.html
McKie, Robin. "Men Redundant? Now We Don't Need Women Either." *The Guardian*. 10 Feb. 2002. Web. 8 Jul. 2016. https://www.theguardian.com/world/2002/feb/10/medicalscience.research
McMillan, Priscilla J. *The Ruin of J. Robert Oppenheimer and the Birth of the Modern Arms Race*. New York: Viking, 2005. Print.
McNicol, Tony. "Japan Sets Sail in Space." *Discover* 26.1 (Jan. 2005): 63. Print.
McStay, Andrew. "Now Advertising Billboards Can Read Your Emotions ... and That's Just the Start." *The Conversation*. 4 Aug. 2015. Web. http://theconversation.com/now-advertising-billboards-can-read-your-emotions-and-thats-just-the-start-45519
Merriam-Webster Dictionary Online. Merriam-Webster, Incorporated. 2015. http://www.merriam-webster.com/
Miller, Walter James, and Frederick Paul Walter, transl. *Jules Verne's 20,000 Leagues under the Sea: The Completely Restored and Annotated Edition*. Annapolis, MD: Naval Institute Press, 1993. Print.
Mitchell, Edward Page. "The Clock That Went Backward." *The Tachypomp and Other Stories*. Seattle: Amazon Digital Services, Inc., 2015. Amazon Kindle.
Mitchell, Edward Page. "The Man Without a Body." *The Tachypomp and Other Stories*. Seattle: Amazon Digital Services, 2015. Amazon Kindle.
Moravec, Hans. *Robot: Mere Machine to Transcendent Mind*. Oxford: Oxford University Press, 1999. Print.
Mori, Masahiro. "The Uncanny Valley." Trans. Karl F. MacDorman and Norri Kageki *IEEE Spectrum*. 12 Jun. 2012. Web. http://spectrum.ieee.org/automaton/robotics/humanoids/the-uncanny-valley
Mosher, Dave. "Astronomers May Now Fully Understand Why the Sky Is Dark at Night." *Business Insider*. 13 Oct. 2016. Web. 14 Oct. 2016. https://amp.businessinsider.com/humble-number-galaxies-space-universe-2016-10
Moskowitz, Sam. ed. *Science Fiction by Gaslight: A History and Anthology of Science Fiction in the Popular Magazines, 1891–1911*. New York: World Publishing, 1968. Print.
"Movie Franchises." *The Numbers—Box Office*

Data, Movie Stars, Idle Speculation. Nash Information Services. 2016. Web. 9 Jul. 2013. http://www.the-numbers.com/movies/franchises/

Muehlhauser, Luke. "What Is AGI?" Machine Intelligence Research Institute. 11 Aug. 2013. Web. 5 Aug. 2016. https://intelligence.org/2013/08/11/what-is-agi/

Munro, John. A Trip to Venus. London: Jarrold & Sons, 1897. Project Gutenberg e-book.

Musk, Elon. "Elon Musk/Full Interview/Code Conference." Youtube. Recode. 2 Jun. 2016. Web. 5 Jun. 2016. https://www.youtube.com/watch?v=wsixsRI-Sz4

Nahin, Paul J. Time Machine: Time Travel in Physics, Metaphysics, and Science Fiction. New York: American Institute of Physics, 1993. Print.

"NASA's Nanosail-D 'Sails' Home—Mission Complete." NASA. Nov. 29, 2011. Web. 15 Jul. 2016. http://www.nasa.gov/mission_pages/smallsats/11-148.html

Nevins, Jess and Michelle Martinez. "Before Science Fiction: Romances of Science and Scientific Romances." io9. 23 Dec. 2011. Web. 15 May 2016. http://io9.gizmodo.com/5870883/science-fiction-before-science-fiction-romances-of-science-and-scientific-romances

"New Issue of the Wendelstein Newsletter." IPP Max-Planck-Institut für Plasmaphysik. Max-Planck-Gesellschaft. 7 Apr. 2016. Web. 15 Apr. 2016. http://www.ipp.mpg.de/40388 17/w7x_letter_12_16

Newitz, Annalee. "Scientists Just Invented the Neural Lace." Gizmodo. 15 Jun. 2015. Web. 1 Jun. 2016. http://gizmodo.com/scientists-just-invented-the-neural-lace-1711540938

Newton, Isaac. A Treatise of the System of the World. Andrew Motte transl. London: F. Fayram, 1728. Print.

Nicholls, Peter. "Big Dumb Objects and Cosmic Enigmas: The Love Affair Between Space Fiction and the Transcendental." gregorybenford.com 2011. Web. http://www.gregorybenford.com/extra/big-dumb-objects-and-cosmic-enigmas/

Nicholls, Peter. The Science in Science Fiction. New York: Alfred A. Knopf, 1983. Print.

Niven, Larry. All the Myriad Ways. New York: Ballantine, 1971. Print.

Niven, Larry, and Jerry Pournelle. Footfall. New York: Del Rey Books, 1985. Print.

Niven, Larry, and Jerry Pournelle. The Mote in God's Eye. Norwalk, CT: Easton Press, 1991. Print.

"The Nobel Prize in Physiology or Medicine 2012." Nobelprize.org. Nobel Media. 8 Oct. 2012. Web. 1 Aug. 2016 http://www.nobelprize.org/nobel_prizes/medicine/laureates/2012/press.html

North, John. The Norton History of Astronomy and Cosmology. New York: W.W. Norton, 1995. Print.

O'Callaghan, Jonathan. "Earth May Have Formed Earlier Than 92% of Other Habitable Planets." IFL Science! 20 Oct. 2015. Web. 16 Sep. 2016. http://www.iflscience.com/space/earth-may-have-formed-earlier-92-other-habitable-planets/

O'Callaghan, Jonathan. "Elon Musk Says We Could Terraform Mars by Dropping Thermonuclear Bombs on It." IFLScience! 10 Sept. 2015. Web. 12 Apr. 2016. http://www.iflscience.com/space/elon-musk-says-we-could-terraform-mars-dropping-thermonuclear-bombs-it

Okuda, Michael, Denise Okuda, and Debbie Mirek. The Star Trek Encyclopedia: A Reference Guide to the Future. New York: Pocket Books, 1994. Print.

Okuda, Michael, and Rick Sternbach. Star Trek: The Next Generation Technical Manual. London: Boxtree, 1991. Print.

Oltion, Jerry. Mudd in Your Eye. New York: Prentice-Hall, 1997. Print.

"One Mission, Two Remarkable Destinations." NASA. Jet Propulsion Laboratory—California Institute of Technology. Web. 27 May 2016. http://dawn.jpl.nasa.gov/science/

"The One Thing We Need to Stop Robots from Achieving World Domination." Forbes. 29 Jan. 2016. Web. 8 Oct. 2016. http://www.forbes.com/sites/quora/2016/01/29/the-one-thing-we-need-to-stop-robots-from-achieving-world-domination/#f4bbfb96dcf0

O'Neill, Timothy R. The Individuated Hobbit: Jung, Tolkien and the Archetypes of Middle-earth. New York: Houghton Mifflin, 1979. Print.

Ophel, Trevor, and John Jenkin. Fire in the Belly: The First Fifty Years of the Pioneer School at the ANU. Canberra: Australian National University, 1996. Web. 7 Aug. 2016. https://physics.anu.edu.au/fire_in_the_belly/Fire_in_the_Belly03.pdf

"Outline History of Nuclear Energy." world-nuclear.org. World Nuclear Association. Mar. 2014. Web. 1 Aug. 2014.

Overbye, Dennis. "Stephen Hawking Joins Russian Entrepreneur's Search for Alien Life." The New York Times. 20 Jul. 2015. Web. 20 Sep. 2016. http://www.nytimes.com/2015/07/21/science/yuri-milner-russian-entrepreneur-promises-100-million-for-alien-search.html?_r=3

Ovid. *Metamorphoses*. Rolfe Humphries transl. Bloomington: Indiana University Press, 1972. Print.

Oxford English Dictionary Online. Ed. John Simpson. Oxford: Oxford University Press, 2014. Web. http://www.oed.com/

Packer, Sharon. *Movies and the Modern Psyche*. Westport, CT: Praeger, 2007. Print.

Palacios-González, César, John Harris, and Giuseppe Testa. "Multiplex Parenting: IVG and the Generations to Come." *Journal of Medical Ethics* March 7 (2014): 1–7. Web. 1 Aug. 2016. http://jme.bmj.com/content/early/2014/03/07/medethics-2013-101810.full

Palazzolo, Joe. "Did the U.S. Make Asteroid Mining Legal?" *The Wall Street Journal*. Dow Jones & Company, Inc. 1 Dec. 2015. Web. 6 Aug. 2016. http://blogs.wsj.com/law/2015/12/01/did-the-u-s-make-asteroid-mining-legal/

Palumbo, Donald E. *The Monomyth in American Science Fiction Films: 28 Visions of the Hero's Journey*. Jefferson, NC: McFarland, 2014. Amazon Kindle.

Pascal, David. "A Brain Is a Terrible Thing to Waste: Mensans, Cryonics, and the Fight to Extend Human Life." *The Cryonics Society*. Mensa Bulletin. Nov./Dec. 2005. Web. 6 Aug. 2016. http://www.cryonicssociety.org/articles_mensajournal.html

Patel, Neel V. "A History of Space Guns from Isaac Newton to Nazis in Paris and Project HARP." Inverse.com. 14 Jun. 2016. Web. 2 Aug. 2016. https://www.inverse.com/article/16735-a-history-of-space-guns-from-isaac-newton-to-nazis-in-paris-and-project-harp

Payne, Robert. *By Me, William Shakespeare: A Biography*. New York: Everest House, 1980. Print.

Pedersen, David L. *Cameral Analysis: A Method of Treating the Psychoneuroses Using Hypnosis*. London: Routledge, 1994. Print.

Pellegrino, Charles. *Flying to Valhalla*. New York: AvoNova Books, 1993, Print.

Pellegrino, Charles, and George Zebrowski. *The Killing Star*. Norwalk, CT: Easton Press, 1995. Print.

Pemberton, T.E. *Freezing a Mother-in-Law*. Boston: Walter H. Baker & Co., 1889. Project Gutenberg e-book.

The Penguin Encyclopedia. Ed. David Crystal. New York: Penguin Books, 2004. Print.

Pickover, Clifford A. *The Physics Book: 250 Milestones in the History of Physics*. New York: Sterling, 2011. Print.

Pilkington, Ace G. "Dog Stars." *Asimov's Science Fiction* 24.5 (2000): 101. Print.

Pilkington, Ace G. "Fighting the History Wars on the Big Screen: From *The Terminator* to *Avatar*." *The Films of James Cameron: Critical Essays*. Ed. Matthew Wilhelm Kapell. Jefferson, NC: McFarland, 2011. Print.

Pilkington, Ace G. "*Forbidden Planet*: Aliens, Monsters and Fictions of Nuclear Disaster." *The Fantastic Made Visible: Essays on the Adaptation of Science Fiction and Fantasy from Page to Screen*. Eds. Matthew Wilhelm Kapell and Ace G. Pilkington. Jefferson, NC: McFarland, 2015. Print.

Pilkington, Ace G. "Introduction: Science Fiction and Fantasy Conquer the World." *The Fantastic Made Visible: Essays on the Adaptation of Science Fiction and Fantasy from Page to Screen*. Eds. Matthew Wilhelm Kapell and Ace G. Pilkington. Jefferson, NC: McFarland, 2015. Print.

Pilkington, Ace G. "*Star Trek*: American Dream, Myth, and Reality." *Star Trek as Myth: Essays on Symbol and Archetype at the Final Frontier*. Ed. Matthew Wilhelm Kapell. Jefferson, NC: McFarland, 2010. Print.

Pilkington, Ace G. "The Taming of the Shrew." *Midsummer Magazine* 1991: 21–23. Print.

Pilkington, Ace G., and Olga A. Pilkington. *Fairy Tales of the Russians and Other Slavs*. Forest Tsar Press, 2009. Print.

Piper, H. Beam. "Omnilingual." *Astounding Science Fiction* Feb. 1957. Print.

Pitzke, Marc. "An International Disaster: Trump Has Shown His True Side, It's Time to Act." *Spiegel Online*. 16 Aug. 2016. Web. 17 Aug. 2016. http://www.spiegel.de/international/world/donald-trump-is-a-threat-to-the-us-and-world-peace-a-1107902.html

Pohl, Frederik. *The Age of the Pussyfoot*. New York: Ballantine Books, 1969. Print.

Pohl, Frederik. "Author's Note." *The Age of the Pussyfoot*. New York: Ballantine Books, 1969. Print.

Pohl, Frederik. "Edgar Rice Burroughs and the Development of Science Fiction." *The Burroughs Bibliophiles*. Web. http://www.burroughsbibliophiles.com/pohl.html

Pollick, Frank E. "In Search of the Uncanny Valley." *Lecture Notes of the Institute for Computer Sciences, Social Informatics and Telecommunications Engineering* 40 (2010): 69–78. Web. http://www.psy.gla.ac.uk/~frank/Documents/InSearchUncannyValley.pdf

Pope-Hennessy, James. *Robert Louis Stevenson*. London: Cassell, 1974. Print.

Popkin, Jeremy D. *From Herodotus to H-Net: The Story of Historiography*. Oxford: Oxford University Press, 2016. Print.

Pournelle, Jerry. "Introduction." *Starswarm*. New York: Baen, 2008. E-book edition.

Powell, Devin. "A Flexible Circuit Has Been Injected Into Living Brains." Smithsonian.com. 8 Jun. 2015. Web. 1 Jun. 2016. http://www.smithsonianmag.com/science-nature/flexible-circuit-has-been-injected-living-brains-180955525/?no-ist

"President Obama Signs Bill Recognizing Asteroid Resource Property Rights into Law." *Planetary Resources.* 25 Nov. 2015. Web. 1 Aug. 2016. http://www.planetaryresources.com/2015/11/president-obama-signs-bill-recognizing-asteroid-resource-property-rights-into-law/

"Press Release: The Nobel Prize in Chemistry 2016." *Nobelprize.org.* Royal Swedish Academy of Sciences. 5 Oct. 2016. Web. 1 Oct. 2016. https://www.nobelprize.org/nobel_prizes/chemistry/laureates/2016/press.html

Priestley, J.B. *The Doomsday Men.* Richmond, VA: Valancourt Books, 2014. Amazon Kindle.

Prucher, Jeff. *Brave New Words: The Oxford Dictionary of Science Fiction.* Oxford: Oxford University Press, 2007. Print.

The Random House Dictionary of the English Language: The Unabridged Edition. Ed. Jess Stein. New York: Random House, 1981. Print.

Random House Unabridged Dictionary: Second Edition. Ed. Stuart Berg Flexner. New York: Random House, 1993. Print.

Rank, Otto. *The Myth of the Birth of the Hero: A Psychological Interpretation of Mythology.* F. Robbins and Smith Ely Jelliffe transl. New York: Journal of Nervous and Mental Disease Publishing Company, 1914. Print.

Ratner, Daniel, and Mark A. Ratner. *Nanotechnology and Homeland Security: New Weapons for New Wars.* Upper Saddle River, NJ: Prentice-Hall, 2004. Print.

Raven, Paul. "Uplift Ethics and Transhuman Hubris." *Futurismic.* 26 Jul. 2011. Web. http://futurismic.com/2011/07/26/uplift-ethics-and-transhuman-hubis/

Reagan, Ronald. *Ronald Reagan: An American Life.* New York: Pocket Books, 1990. Print.

Regis, Ed. *Monsters: The Hindenburg Disaster and the Birth of Pathological Technology.* New York: Basic Books, 2015. Print.

Regis, Ed. *Nano, the Emerging Science of Nanotechnology: Remaking the World—Molecule by Molecule.* Boston: Little, Brown, 1995. Print.

Regis, Ed. "What Could Go Wrong?" *Slate.* Slate Group. 30 Sept. 2015. Web. 5 March 2016. http://www.slate.com/articles/technology/future_tense/2015/09/project_plowshare_the_1950s_plan_to_use_nukes_to_make_roads_and_redirect.html

Reichardt, Jasia. *Robots. Fact, Fiction, and Prediction.* New York: Penguin Books, 1978. Print.

Reichhardt, Tony. "The First Countdown?" *Air & Space Smithsonian.* 26 Feb. 2011. Web. http://blogs.airspacemag.com/daily-planet/2011/02/the-first-countdown/

Reoma, Junewai L., et al. "Development of an Artificial Placenta I: Pumpless Arterio-venous Extracorporeal Life Support in a Neonatal Sheep Model." *Journal of Pediatric Surgery* 44 (2009): 53–59. Print.

Rhodes, Richard. *The Making of the Atomic Bomb.* New York: Simon & Schuster, 2012. Amazon Kindle.

Ribner, Irving. "Preface." *Christopher Marlowe's Doctor Faustus: Text and Major Criticism.* Ed. Irving Ribner. New York: Odyssey Press, 1966. Print.

Rice, Jordan. "The First Private Spaceflight Company is Cleared for a Moon Landing." Astronomy.com. Kalmbach Publishing Co. 4 Aug. 2016. Web. 4 Sep. 2016. http://www.astronomy.com/news/2016/08/next-stop-the-moon

Ringo, John. *There Will Be Dragons.* New York: Baen, 2003. Baen e-book.

Roberts, Adam. *The History of Science Fiction.* New York: Palgrave Macmillan, 2005. Print.

Roberts, Andrew. *Napoleon: A Life.* New York: Viking, 2014. Print.

Roberts, J.M. *Triumph of the West.* British Broadcasting Corporation, 1985. Film.

Robinson, Spider. *Lifehouse.* Wake Forest, NC: Baen Books, 1997. Print.

Rogow, Roberta. *Futurespeak: A Fan's Guide to the Language of Science Fiction.* New York: Paragon House, 1991. Print.

Rohn, Jennifer. "What Is Lab Lit (the Genre)?" lablit.com. Jennifer Rohn. 7 Mar. 2005. Web. 14 Feb. 2016. http://www.lablit.com/article/3

Romaine, Suzanne. *The Cambridge History of the English Language. Volume IV: 1776–1997.* Cambridge: Cambridge University Press, 1998. Print.

Rose, Steve. "Constructive Criticism: the Week in Architecture." *The Guardian* 8 June 2012. Web. http://www.theguardian.com/artanddesign/2012/jun/08/week-architecture-ray-bradbury-curtain

Rose, Thomas. "Going Ape over Human Rights." *CBC News.* 2 Aug. 2007. Web. 7 Oct. 2015. https://web.archive.org/web/20100203225450/http://www.cbc.ca/news/viewpoint/vp_rose/20070802.html

Rowland, Marcus L. "Notes from the Editor." *Stories of Other Worlds and A Honeymoon in Space.* George Griffith. Somerville, MA: Heliograph, 2000. Print.

Rottensteiner, Franz, ed. *The Black Mirror and Other Stories: An Anthology of Science Fiction from Germany and Austria*. Middletown, CT: Wesleyan University Press, 2008. Print.
Rucker, Rudy, and Bruce Sterling. "Totem Poles." Tor.com. Macmillan. 10 Aug. 2016. Web. 11 Aug. 2016. http://www.tor.com/2016/08/10/totem-poles/?utm_source= exacttarget&utm_medium=newsletter& utm_term=tordotcom-tordotcomnews letter&utm_content=na-readblog-blogpost &utm_campaign=9780765385017
Russell, A. Kingsley. "A Golden Era of Science Fiction and Fantasy." *Science Fiction by the Rivals of H.G. Wells*. Secaucus, NJ: Castle Books, 1979. Print.
Russell, Eric Frank. "Mana." *The Best of Eric Frank Russell*. New York: Ballantine, 1978. Print.
Russell, Jeffrey Burton. *Witchcraft in the Middle Ages*. Secaucus, NJ: Citadel, 1972. Print.
Russon, Mary-Ann. "Neural Lace has been Invented to Organically Connect Your Brain with a Computer." *International Business Times*. IBTimes Co., Ltd. 16 Jun. 2015. Web. 14 May 2016. http://www.ibtimes.co.uk/neural-lace-has-been-invented-organically-connect-your-brain-computer-1506481
Rutherford, Ernest. *Radio-Activity*. Cambridge: Cambridge University Press, 1905. Print.
Ryan, Cornelius. "Introduction." *Across the Space Frontier*. Ed. Cornelius Ryan. New York: Viking, 1952. Print.
Saberhagen, Fred. "Starsong." *World's Best Science Fiction 1969*. Eds. Donald A. Wollheim and Terry Carr. New York: Ace Books, 1969. Print.
Saberhagen, Fred. "Without a Thought." *Berserker*. Albuquerque: JSS Literary Productions, 1967. Amazon Kindle.
Sagan, Carl. *Contact*. New York: Simon & Schuster, 1985. Print.
Sagan, Carl. *Cosmos*. New York: Random House, 1980. Print.
"Samsung 60" Class 4K Ultra HD Smart SUHD TV-UN60KS8000." *Kmart*. 2016. Web. 1 Nov. 2016. http://www.kmart.com/samsung-60inch-class-4k-ultra-hd-smart-suhd/p-05772018000P?plpSellerId=Sears&prdNo= 1&blockNo=1&blockType=G1
Saunders, Russell. "Clipper Ships of Space." *Astounding Science Fiction* May 1951: 136–143. Print.
Sawyer, Robert J. "Just Like Old Times." *Dinosaur Fantastic*. Eds. Mike Resnick and Martin H. Greenberg. New York: DAW, 1993. Print.
Sawyer, Robert J. "You See But You Do Not Observe." SFWriter.com. 1995. Web. 5 Aug. 2016. http://sfwriter.com/styousee.htm
Schmundt, Hilmar, and Phil Thoma. "Touring Tragedy: A Day of Disaster Porn in Chernobyl." *Spiegel Online*. Der Spiegel. 25 Apr. 2016. Web. 28 Apr. 2016. http://www.spiegel.de/international/world/tourism-is-booming-in-the-chernobyl-exclusion-zone-a-1089210.html
Searles, A. Langley. "Introduction." *Edison's Conquest of Mars*. Garrett P. Serviss. Los Angeles: Carcosa House, 1947. http://www.gutenberg.org/ebooks/19141
Segen, J.C. *The Dictionary of Modern Medicine*. New York: Parthenon, 1992. Print.
Sellars, Peter. "Interview with Peter Sellars." *Doctor Atomic*. Dir. Peter Sellars. Opus Arte, 2008. DVD/Opera.
Serviss, Garrett P. *A Columbus of Space*. Project Gutenberg e-book.
SETI Institute. "Fermi Paradox." *SETI.org*. SETI Institute. 2016. Web. 20 Sep. 2016. http://www.seti.org/seti-institute/project/details/fermi-paradox
Shakespeare, William. *Hamlet*. *The Norton Shakespeare*. Ed. Stephen Greenblatt. New York: W.W. Norton, 1997. Print.
Shakespeare, William. *Romeo and Juliet*. *The Norton Shakespeare*. Ed. Stephen Greenblatt. New York: W.W. Norton, 1997. Print.
Shakespeare, William. *The Tempest*. *The Norton Shakespeare*. Ed. Stephen Greenblatt. London: W.W. Norton, 1997. Print.
Shakespeare, William. *The Winter's Tale*. *The Norton Shakespeare*. Ed. Stephen Greenblatt. New York: W.W. Norton, 1997. Print.
Shatner, William. Narrator. *How William Shatner Changed the World*. Allumination Filmworks. 2005. Film.
Shaw, Bernard. *As Far As Thought Can Reach*. In *Complete Plays with Prefaces*, vol. II. New York: Dodd, Mead, 1963. Print.
Sheffield, Charles. "The Birth and Death of Science Fiction." Dixie State College, St. George, UT. 25 March 1998. Public lecture.
Sheffield, Charles. *The Borderlands of Science: How to Think Like a Scientist and Write Science Fiction*. New York: Baen, 1999. Print.
Sheffield, Charles. *Cold as Ice*. New York: Tom Doherty Associates, 1992. Print.
Sheffield, Charles. *Earth Watch: A Survey of the World from Space*. New York: Macmillan, 1981. Print.
Sheffield, Charles. "Skystalk." *Vectors*. New York: Ace Books, 1979. Print.
Sheffield, Charles. *Tomorrow and Tomorrow*. Norwalk, CT: Easton Press, 1997. Print.
Shelley, Percy Bysshe. "Ozymandias." *The Com-*

plete *Poetical Works of Percy Bysshe Shelley*. Ed. Thomas Hutchinson. London: Oxford University Press, 1914. Print.

Shumaker, Wayne. *The Occult Sciences in the Renaissance: A Study in Intellectual Patterns*. Berkeley: University of California Press, 1972. Print.

Silverberg, Robert. *Gilgamesh the King*. New York: Open Road Media, 2013. Amazon Kindle.

Simak, Clifford D. *City*. Norwalk, CT: Easton Press, 1995. Print.

Simak, Clifford D. *A Heritage of Stars*. New York: Berkley, 1977. Print.

Simak, Clifford D. "New Folks' Home." *Best Science Fiction Stories of Clifford D. Simak*. New York: Doubleday, 1965. Print.

Simak, Clifford D. *Skirmish: The Great Short Fiction of Clifford D. Simak*. New York: G. P. Putnam's Sons, 1977. Print.

Simak, Clifford D. *Way Station*. New York: Doubleday, 1963. Print.

Simak, Clifford D. *Why Call Them Back from Heaven?* New York: Avon, 1967. Print.

Simberg, Rand. "What Ion Propulsion Means for Boeing—And Our Future in Space." *Popular Mechanics* 28 Mar. 2012. Web. 10 Apr 2016. http://www.popularmechanics.com/space/rockets/a7513/what-ion-propulsion-means-for-boeing-and-our-future-in-space-7685623/

Singer, P.W., and August Cole. *Ghost Fleet*. New York: Eamon Dolan, 2015. Print.

"Smart 1." *European Space Agency*. 21 Aug. 2009. Web. 18 May 2016. http://www.esa.int/Our_Activities/Space_Science/SMART-12

Smith, Cordwainer. "Alpha Ralpha Boulevard." *We the Underpeople*. New York: Baen, 2006. Print.

Smith, Cordwainer. "The Lady Who Sailed *The Soul*." *The Best of Cordwainer Smith*. Ed. J.J. Pierce. New York: Nelson Doubleday, 1975. Print.

Smith, P.D. *Doomsday Men: The Real Dr. Strangelove and the Dream of the Superweapon*. New York: St. Martin's Press, 2007. Print.

Soddy, Frederick. *Wealth, Virtual Wealth and Debt*. New York: E. P. Dutton, 1933. Print.

Sofge, Erik. "The Truth about Robots and the Uncanny Valley: Analysis." *Popular Mechanics*. 20 Jan. 2010. Web. 5 Nov. 2015. http://www.popularmechanics.com/technology/robots/a5001/4343054/

"Solar Sail Stunner." *NASA Science/Science News*. NASA. 24 Jan. 2011. Web. 17 Jun. 2016. http://science.nasa.gov/science-news/science-at-nasa/2011/24jan_solarsail/

Sontag, Susan. *Against Interpretation and Other Essays*. New York: Farrar, Straus & Giroux, 1966. Print.

Stableford, Brian. *Historical Dictionary of Science Fiction Literature*. Lanham, MD: Scarecrow Press, 2004. Print.

Stableford, Brian. *Science Fact and Science Fiction: An Encyclopedia*. New York: Routledge, 2006. Print.

Stapledon, Olaf. *Odd John and Sirius*. New York: Dover, 1972. Print.

Statt, Nick. "Scientists Achieve Reliable Quantum Teleportation for First Time." *Cnet*. 29 May 2014. Web. 16 June 2016. http://www.cnet.com/news/scientists-achieve-reliable-quantum-teleportation-for-the-first-time/

Sterling, Bruce. "Homo Sapiens Declared Extinct." *Supermen: Tales of the Posthuman Future*. Ed. Gardner Dozois. New York: St. Martin's Griffin, 2002. Print.

Sterling, Bruce, and Lewis Shiner. "Mozart in Mirrorshades." Ed. Victoria Blake. *Cyberpunk: Stories of Hardware, Software, Wetware, Evolution and Revolution*. Portland, OR: Underland Press. Print.

Stevenson, Robert Louis. *The Strange Case of Dr. Jekyll and Mr. Hyde*. New York: Barnes and Noble, 1995. Print.

Strugatsky, Arkady, and Boris Strugatsky. *Roadside Picnic*. Chicago: Chicago Review Press, 2012. Print.

Suter, Sonia M. "In Vitro Gametogenesis: Just Another Way to Have a Baby?" *Journal of Law and the Biosciences* December 17 (2015): 1–33. Web. 25 Jul. 2016. http://jlb.oxfordjournals.org/content/early/2015/12/16/jlb.lsv057.full

Svitil, Kathy A. "Teleportation Gets Real." *Discover* 26.1 (Jan. 2005): 72. Print.

Talbot, Margaret. "Pixel Perfect: The Scientist behind the Digital Cloning of Actors." *The New Yorker*. 28 Apr. 2014. Web. http://www.newyorker.com/magazine/2014/04/28/pixel-perfect-2

Taves, Brian, and Stephen Michaluk, eds. *The Jules Verne Encyclopedia*. Lanham, MD: Scarecrow Press, 1996. Print.

"Temporal Shock Wave." *Memory Alpha*. Web. 1 Jun. 2016. http://memory-alpha.wikia.com/wiki/Temporal_shock_wave

Tenn, William. "Bernie the Faust." *Bookzz.org*. Web/PDF. 6 May 2016. http://bookzz.org/g/Tenn%20William

Tenn, William. "Venus and the Seven Sexes." *BookReader*. 2016. Web. 2 Aug. 2016. http://bookre.org/reader?file=289659&pg=24

Tennyson, Alfred. "Locksley Hall." *Harvard Classics: English Poetry, Vol. III From Tennyson to Whitman*. New York: P.F. Collier and Son,

1914. Web. 10 Jun. 2016. http://www.bartleby.com/42/636.html

Thorne, Kip S. *Black Holes & Time Warps: Einstein's Outrageous Legacy*. New York: W.W. Norton, 1994. Print.

Tipler, Frank J. *The Physics of Immortality*. New York: Doubleday, 1994. Print.

Tolkien, J.R.R. "On Fairy-Stories." *The Tolkien Reader*. New York: Ballantine Books, 1966. Print.

Tolstoy, Alexei. *Engineer Garin and His Death Ray*. Moscow: Raduga Publishers, 1987. Print.

Tsiolkovsky, Konstantin E. *Selected Works of Konstantin E. Tsiolkovsky*. Honolulu: University Press of the Pacific, 2004. Print.

Tucker, Patrick. "Can the Navy's Electric Cannon Be Saved?" Defenseone.com. 2 Jun. 2016. Web. 8 Sep. 2016. http://www.defenseone.com/technology/2016/06/can-navys-electric-cannon-be-saved/128793/

Turing, A.M. "Computing Machinery and Intelligence." *Mind* 59 (1950): 433–460. Web. 2 Jun. 2016. http://www.loebner.net/Prizef/TuringArticle.html

Turnbow, Gene. "DARPA Working on Sentient Robots With 'Positronic' Brains." *Krypton Radio*. Krypton Media Group. 17 Jul. 2013. Web. 2 Jul. 2016. https://kryptonradio.com/2013/07/17/darpa-working-on-sentient-robots-with-positronic-robot-brains/

"Turning Up The Heat: Latest Evolution Of Lockheed Martin Laser Weapon System Stops Truck In Field Test." Lockheedmartin.com. Mar. 2015. Web. 5 Aug. 2016. http://lockheedmartin.com/us/news/press-releases/2015/march/ssc-space-athena-laser.html

Turtledove, Harry. *Agent of Byzantium*. Riverdale, NY: Baen, 1994. Print.

Turtledove, Harry. "Introduction." *The Arrows of Hercules*. L. Sprague de Camp. Rockville, MD: Phoenix Pick, 2014. Amazon Kindle.

Twain, Mark. *A Connecticut Yankee in King Arthur's Court*. New York: Nelson Doubleday, n.d. Print.

Twain, Mark. "From the 'London Times' of 1904." *Readbookonline.net*. Web. 20 Jun. 2015. http://www.readbookonline.net/readOnLine/998/

Twilley, Nicola. "Artificial Intelligence Goes to the Arcade." *The New Yorker* 25 Feb. 2015. Web. 27 May 2016. http://www.newyorker.com/tech/elements/deepmind-artificial-intelligence-video-games?curator=MediaREDEF

"Uplift." *Encyclopedia of Science Fiction, The*. 2014. Web. http://www.sf-encyclopedia.com/

Urban, Tim. "The Fermi Paradox." *Wait But Why*. May 2014. Web. 1 May 2016. http://waitbutwhy.com/2014/05/fermi-paradox.html

Uttley, Alison. *A Traveller in Time*. New York: New York Review Children's Collection, 1939. Print.

Van Vogt, A. E. "Asylum." *Adventures in Time and Space: An Anthology of Science Fiction Stories*. Eds. Raymond J. Healey and J. Francis McComas. New York: Random House, 1957. Print.

Van Vogt, A. E. "Research Alpha." *More Than Superhuman*. New York: Dell, 1971. Print.

Varley, John. "Air Raid." *The John Varley Reader*. New York: Ace Books, 2004. Amazon Kindle.

Verne, Jules. "Cable from Jules Verne to Simon Lake." *Simon Lake Submarine Web Site*. Web. 5 Aug. 2015. http://www.simonlake.com/html/jules_verne.html

Verne, Jules. *The Castle of the Carpathians with a Critical Introduction by Ace G. Pilkington*. Forest Tsar Press, 2010. Print.

Verne, Jules. *From the Earth to the Moon*. New York: Baen, 2013. Baen e-book.

Verne, Jules. *From the Earth to the Moon, Direct in Ninety-Seven Hours and Twenty Minutes: and a Trip Round It*. New York: Scribner, Armstrong & Company, 1874. Project Gutenberg e-book.

Verne, Jules. *The Mysterious Island*. Middletown, CT: Wesleyan University Press, 2001. Print.

Vinge, Vernor. *Marooned in Realtime*. New York: Baen, 1986. Print.

Vinge, Vernor. *The Peace War*. New York: Tor Books, 2007. Amazon Kindle.

von Braun, Wernher. "Prelude to Space Travel." *Across the Space Frontier*. Ed. Cornelius Ryan. New York: Viking, 1952. Print.

Von Hippel. Arthur. "Molecular Designing of Materials." *Science* 138.3537 (1962): 91–108. Print.

Von Hippel, Arthur. *Molecular Science and Molecular Engineering*. Cambridge, MA: MIT Press, 1959. Print.

Waldman, Harry. *Dictionary of Robotics*. New York: Macmillan, 1985. Print.

Waldman, Katy. "The Nuclear Monsters That Terrorized the 1950s." *Slate*. 31 Jan. 2013. Web. 2 Aug. 2014. http://www.slate.com/articles/health_and_science/nuclear_power/2013/01/nuclear_monster_movies_sci_fi_films_in_the_1950s_were_terrifying_escapism.html

Walter, William J. *Space Age*. New York: Random House, 1992. Print.

Waltonen, Karma. "Loving the Other in Science Fiction by Women." *MOSF Journal of Science*

Fiction 1.1 (2016): 33–44. Web. 4 May 2016. http://publish.lib.umd.edu/scifi/article/view/250

Warrick, Patricia, Martin Harry Greenberg, and Joseph Olander, eds. *Science Fiction: Contemporary Mythology*. New York: Harper & Row, 1978. Print.

Watson, Sara M. "Data Doppelgängers and the Uncanny Valley of Personalization." *The Atlantic*. 16 Jun. 2014. Web. 7 Oct. 2014. http://www.theatlantic.com/technology/archive/2014/06/data-doppelgangers-and-the-uncanny-valley-of-personalization/372780/

Watson, Stephanie. "How Cryonics Works." *How Stuff Works: Science*. InfoSpace Holdings. 2016. Web. 10 Aug. 2016. http://science.howstuffworks.com/life/genetic/cryonics.htm/printable

Watters, Thomas R. *Planets: A Smithsonian Guide*. New York: Macmillan, 1995.

Watterson, Bill. *Scientific Progress Goes "Boink."* Kansas City: Andrews and McMeel, 1991. Print.

Watts, Peter. "Margaret Atwood and the Hierarchy of Contempt." Rifters.com. PDF. 14 Aug. 2016. http://www.rifters.com/real/shorts/PeterWatts_Atwood.pdf

Weber, David. "Biography." *David Weber*. 2016. Web. 4 Nov. 2016. http://www.davidweber.net/biography

Weber, David. *Out of the Dark*. New York: Tor, 2011. Print.

Weinbaum, Stanley G. *A Martian Odyssey*. New York: Lancer Books, 1972. Print.

Wells, H. G. *The Island of Doctor Moreau*. New York: Penguin Classics, 2005. Print.

Wells, H. G. *The Last War: A World Set Free*. Lincoln: University of Nebraska Press/Bison Books, 2001. Print.

Wells, H. G. "The Limits of Individual Plasticity." *H.G. Wells: Early Writings in Science and Science Fiction*. Eds. Robert Philmus and David Y. Hughes. Berkeley: University of California Press, 1975. Print.

Wells, H. G. *Men Like Gods*. Mt. Pleasant, SC: Arcadia Publishing, 2016. Amazon Kindle.

Wells, H. G. "Preface." *The Complete Science Fiction Treasury of H.G. Wells*. New York: Avenel Books, 1978 (first published by Alfred A. Knopf in 1934). Print.

Wells, H. G. *The Time Machine: An Invention: A Critical Text of the 1895 London First Edition, with an Introduction and Appendices*. Ed. Leon Stover. Jefferson, NC: McFarland, 2012. Print.

Wells, H. G. *Tono-Bungay*. New York: Modern Library, 1935. Print.

Wells, H. G. *The War of the Worlds*. New York: Tribeca Publishing, 2013. Print.

Wenz, John. "NASA's Next Big Telescope Draws One Step Closer to the Stars." *Popular Mechanics*. 3 Feb. 2016. Web. 17 Jun. 2016. http://www.popularmechanics.com/space/telescopes/news/g2455/nasas-next-big-telescope-draws-one-step-closer-to-the-stars/

"Who Did Actually Invent the Word 'Robot' and What Does It Mean?" capek.misto.cz. Web. 13 Feb. 2015. http://capek.misto.cz/english/robot.html

Whoriskey, Peter. "What Actual 'Caveman' DNA Says About the Paleo Movement." *The Washington Post*. 17 Feb. 2016. Web. https://www.washingtonpost.com/news/wonk/wp/2016/02/17/what-actual-caveman-dna-says-about-the-paleo-movement/?tid=a_inl

Wilde, Fran. "How Do You Like Your Science Fiction? Ten Authors Weigh In on 'Hard' vs. 'Soft' SF." *Tor*. 21 Jan. 2016. Web. 17 Oct. 2016. http://www.tor.com/2016/01/21/how-do-you-like-your-science-fiction-ten-authors-weigh-in-on-hard-vs-soft-sf/

Williams, Matt. "Earth-Like Planet Around Proxima Centauri Discovered." *Universe Today*. 13 Aug. 2016. Web. 16 Aug. 2016. http://www.universetoday.com/130276/earth-like-planet-around-proxima-centauri-discovered/

Wilson, Taylor. "My Radical Plan for Small Nuclear Fission Reactors." Transcribed by Joseph Geni. TED. TED Conferences, LLC. Apr. 2013. Web. 1 Mar. 2016. http://www.ted.com/talks/taylor_wilson_my_radical_plan_for_small_nuclear_fission_reactors/transcript?language=en

Wixon, David W. "Introduction." *Good Night, Mr. James: And Other Stories (The Complete Short Fiction of Clifford D. Simak, Volume 8)*. New York: Open Road Media, 2016. Amazon Kindle.

Wodehouse, P.G. *Spring Fever*. London: Penguin Books, 1969. Print.

Wolchover, Natalie. "Why Do Americans and Brits Have Different Accents?" *Live Science*. 9 Jan. 2012. Web. http://www.livescience.com/33652-americans-brits-accents.html

Wollheim, Donald A. *The Universe Makers*. New York: Harper & Row, 1971. Print.

Woodman, David. *White Magic and English Renaissance Drama*. Rutherford, NJ: Fairleigh Dickinson University Press, 1973. Print.

Wu, Dingbo. "Looking Backward: An Introduction to Chinese Science Fiction." *Science Fiction from China*. Ed. Dingbo Wu and Patrick D. Murphy. New York: Praeger, 1989. Print.

Yefremov, Ivan. "The Heart of the Serpent." *More Soviet Science Fiction*. Ed. Isaac Asimov. Toronto: Collier Books, 1962. Print.

Yonas, Gerold, and Jill Gibson. "Sword of Heat." *STEPS*. Potomac Institute for Policy Studies. 29 Jun. 2015. Web. 29 Jul. 2016. http://www.potomacinstitute.org/steps/featured-articles/15-sword-of-heat

Zahn, Timothy. *A Coming of Age*. New York: Open Road Media, 2012. Print.

Zarkadakis, George. *In Our Own Image*. New York: Pegasus Books, 2015. Print.

Zebrowski, George. "Afterword: Reality Check." Charles Pellegrino and George Zebrowski. *The Killing Star*. Norwalk, CT: Easton Press, 1995. Print.

Zimmermann, Nils. "Nuclear Accidents Make Mutant Bugs and Birds." *DW*. Deutsche Welle. 26 Apr. 2016. Web. 28 Apr. 2016. http://www.dw.com/en/nuclear-accidents-make-mutant-bugs-and-birds/a-19098683

Zipes, Jack. *Spells of Enchantment: The Wondrous Fairy Tales of Western Culture*. New York: Viking, 1991. Print.

Zubrin, Robert. *Entering Space: Creating a Spacefaring Civilization*. New York: Jeremy P. Tarcher/Putnam, 1999. Print.

Zuvela, Matt. "Germany Throws the Switch on Seven Aging Nuclear Reactors." *DW*. Deutsche Welle. 19 March 2011. Web. 8 Feb. 2016. http://www.dw.com/en/germany-throws-the-switch-on-seven-aging-nuclear-reactors/a-14924547

Film and Television

The Abyss. Dir. James Cameron. Twentieth Century–Fox, 1989. Film.

Alien. Dir. Ridley Scott. Brandywine Productions, Twentieth Century–Fox, 1979. Film.

Aliens. Dir. James Cameron. Twentieth Century–Fox, 1986. Film.

Andromeda. "The Music of a Distant Drum." Season 1, episode 13. Dir. Allan Kroeker. BLT Productions, 5 February 2001. Television.

Andromeda. "The Honey Offering." Season 1, episode 19. Dir. Brad Turner. BLT Productions, 23 Apr. 2001. Television.

Andromeda. "Una Salus Victus." Season 2, episode 7. Dir. Allan Kroeker. BLT Productions, 12 Nov. 2001. Television.

Andromeda. "What Will Be Was Not." Season 5, episode 9. Dir. Gordon Verheul. BLT Productions, 19 Nov. 2004. Television.

Avatar. Dir. James Cameron. Twentieth Century–Fox, 2009. Film.

Avengers, The. Dir. Joss Whedon. Marvel Studios, 2012. Film.

Babylon 5: A Call to Arms. Dir. Michael Vejar. Babylonian Productions, 1999. Film.

Babylon 5: In the Beginning. Dir. Michael Vejar. Babylonian Productions, 1998. Film.

Babylon 5. "Phoenix Rising." Season 5, episode 11. Dir. David J. Eagle. Babylonian Productions, 1 Apr. 1998. Television.

Babylon 5. "Whatever Happened to Mr. Garibaldi?" Season 4, episode 2. Dir. Kevin James Dobson. Babylonian Productions, 11 Nov. 1996. Television.

The Beautician and the Beast. Dir. Ken Kwapis. High School Sweethearts, 1997. Film.

Beneath the Planet of the Apes. Dir. Ted Post. Twentieth Century–Fox, 1970. Film.

Berkeley Square. Dir. Frank Lloyd. Fox Film Corporation, 1933. Film.

"Boeing TV Spot, 'You Just Wait.'" *iSpot.tv*. Web. 25 Jul. 2016. Video/Commercial. https://www.ispot.tv/ad/Attf/boeing-you-just-wait

Childhood's End. Dir. Nick Hurran. Universal Studios, 2015. Television.

The China Syndrome. Dir. James Bridges. Columbia Pictures, 1979. Film.

Columbo. "Mind Over Mayhem." Season 3, episode 6. Dir. Alf Kjellin. Universal Television, 10 February 1974. Television.

Countdown. Dir. Robert Altman. William Conrad Productions, 1967. Film.

Demolition Man. Dir. Marco Brambilla. Warner Bros., 1993. Film.

Dr. Strangelove or How I Learned to Stop Worrying and Love the Bomb. Dir. Stanley Kubrick. Columbia Pictures, 1964. Film.

Doctor Who. "A Good Man Goes to War." Season 6, episode 7. Dir. Peter Hoar. BBC Wales, 4 Jun. 2011. Television.

Doctor Who. "The Parting of the Ways." Season 1, episode 13. Dir. Joe Ahearne. BBC Wales, 9 June 2006. Television.

Extant. "Re-Entry." Season 1, episode 1. Dir. Allen Coulter. 22 Plates, 9 July 2014. Television.

Forbidden Planet. Dir. Fred M. Wilcox. MGM, 1956. Film.

GoldenEye. Dir. Martin Campbell. Eon Productions, 1995. Film.

How William Shatner Changed the World. Dir. Julian Jones. Allumination Filmworks, 2005.

Iceman. Dir. Fred Schepisi. Universal Pictures, 1984. Film.

Independence Day. Dir. Roland Emmerich. Twentieth Century–Fox, 1996. Film.

Interstellar. Dir. Christopher Nolan. Paramount Pictures, 2014. Film.

The Invisible Boy. Dir. Herman Hoffman. MGM, 1957. Film.

Logan's Run. "Logan's Run." Season 1, episode 1. Dir. Robert Day. MGM, 16 Sep. 1977. Television.

Film and Television

Lost in Space: "The Wreck of the Robot." Season 2, episode 13. Dir. Nathan Juran. Irwin Allen Productions, 14 Dec. 1966. Television.
MacGyver. Henry Winkler/John Rich Productions, 1985–1992. Television.
Millennium. Dir. Michael Anderson. First Millenium Partnership, 1989. Film.
The Mothman Prophecies. Dir. Mark Pellington. Lakeshore Entertainment, 2002. Film.
The Neanderthal Man. Dir. E.A. Dupont. Global Productions, 1953. Film.
Oz the Great and Powerful. Dir. Sam Raimi. Walt Disney Studios Motion Pictures, 2013.
Person of Interest. "Bad Code." Season 2, episode 2. Dir. Jon Cassar. Kilter Films, 4 Oct. 2012. Television.
Person of Interest. "The Cold War." Season 4, episode 10. Dir. Michael Offer. Kilter Films, 16 Dec. 2014. Television.
Person of Interest. "The Contingency." Season 2, episode 1. Dir. Richard J. Lewis. Kilter Films, 27 Sept. 2012. Television.
Person of Interest. "Prophets." Season 4, episode 5. Dir. Kenneth Fink. Kilter Films, 21 Oct. 2014. Television.
Person of Interest. "Synecdoche." Season 5, episode 11. Dir. Tim Matheson. Kilter Films, 7 Jun. 2016. Television.
Person of Interest. "Triggerman." Season 2, episode 4. Dir. James Whitmore, Jr. Kilter Films, 25 Oct. 2012. Television.
Person of Interest. "Wolf and Cub." Season 1, episode 14. Dir. Chris Fisher. Kilter Films, 9 Feb. 2012. Television.
Planet of the Apes. Dir. Franklin J. Schaffner. Twentieth Century–Fox, 1968. Film.
Real Time with Bill Maher. "Episode 386." HBO, 29 Apr. 2016. Talk Show.
Romanoff and Juliet. Dir. Peter Ustinov. Universal Pictures, 1960. Film.
Stargate: Atlantis. "The Siege (Part 2)." Season 1, episode 19. Dir. Martin Wood. Acme Shark, 25 Mar. 2005. Television.
Stargate SG-1. "Children of the Gods." Season 1, episode 1. Dir. Mario Azzopardi. Double Secret Productions, 27 July 1997. Television.
Star Trek. "The Apple." Season 2, episode 5. Dir. Joseph Pevney. Desilu Productions, 13 Oct. 1967. Television.
Star Trek. "Bread and Circuses." Season 2, episode 25. Dir. Ralph Senensky. Paramount, 15 Mar. 1968. Television.
Star Trek. "The Devil in the Dark." Season 1, episode 25. Dir. Joseph Pevney. Desilu Productions, 9 Mar. 1967. Television.
Star Trek. "The Doomsday Machine." Season 2, episode 6. Dir. Marc Daniels. Desilu Productions, 20 Oct. 1967. Television.
Star Trek. "The Enemy Within." Season 1, episode 5. Dir. Leo Penn. Desilu Productions, 6 Oct. 1966. Television.
Star Trek. "Spock's Brain." Season 3, episode 1. Dir. Marc Daniels. Paramount, 1968. Television.
Star Trek. "The Ultimate Computer." Season 2, episode 24. Dir. John Meredyth Lucas. Paramount, 8 Mar. 1968. Television.
Star Trek. "Who Mourns for Adonais?" Season 2, episode 2. Dir. Marc Daniels. Desilu Productions, 22 Sep. 1967. Television.
Star Trek: First Contact. Dir. Jonathan Frakes. Paramount, 1996. Film.
Star Trek Into Darkness. Dir. J. J. Abrams. Paramount, 16 May 2013. Film.
Star Trek: The Next Generation. "Datalore." Season 1, episode 12. Dir. Rob Bowman. Paramount, 16 Jan. 1988. Television.
Star Trek: The Next Generation. "Descent Part 1." Season 6, episode 26. Dir. Alexander Singer. Paramount, 19 June 1993. Television.
Star Trek: The Next Generation. "Emergence." Season 7, episode 23. Dir. Cliff Bole. Paramount, 7 May 1994. Television.
Star Trek: The Next Generation. "The Measure of a Man." Season 2, episode 9. Dir. Robert Scheerer. Paramount, 11 Feb. 1992. Television.
Star Trek: The Next Generation. "The Neutral Zone." Season 1, episode 25. Dir. James L. Conway. Paramount, 16 May 1988. Television.
Star Trek: The Next Generation. "Realm of Fear." Season 6, episode 2. Dir. Cliff Bole. Paramount, 26 Sep. 1992. Television.
Star Trek: The Next Generation. "Second Chances." Season 6, episode 24. Dir. LeVar Burton. Paramount, 22 May 1993. Television.
Star Trek II: The Wrath of Kahn. Dir. Nicholas Meyer. Paramount, 1982. Film.
Star Trek: Voyager. "The Gift." Season 4, episode 2. Dir. Anson Williams. Paramount, 10 Sept. 1997. Television.
Star Trek: Voyager. "Infinite Regress." Season 5, episode 7. Dir. David Livingston. Paramount, 25 Nov. 1998. Television.
Star Trek: Voyager. "Prey." Season 4, episode 16. Dir. Allan Eastman. Paramount, 18 Feb. 1998. Television.
Star Trek: Voyager. "Scientific Method." Season 4, episode 7. Dir. David Livingston. Paramount, 29 Oct. 1997. Television.
Star Trek: Voyager. "Year of Hell." Part I. Season 4, episode 8. Dir. Allan Kroeker. Paramount, 5 Nov. 1997. Television.
Star Trek: Voyager. "Year of Hell." Part II. Season 4, episode 9. Dir. Mike Vejar. Paramount, 12 Nov. 1997. Television.

Works Cited

Film and Television

Star Wars: Episode IV—A New Hope. Dir. George Lucas. Lucasfilm, 1977. Film.

Star Wars: Episode VII—The Force Awakens. Dir. J. J. Abrams. Lucasfilm, 2015. Film.

The Stepford Wives. Dir. Bryan Forbes. Columbia Pictures, 1975. Film.

Superman III. Dir. Richard Lester. Dovemead Films, 1983. Film.

The Terminator. Dir. James Cameron. Hemdale, 1984. Film.

Terminator Genisys. Dir. Alan Taylor. Paramount, 2015. Film.

Terminator 2: Judgment Day. Dir. James Cameron. Carolco Pictures, 1991. Film.

Terra Nova. "Genesis: Part 1." Season 1, episode 1. Dir. Alex Graves. Amblin, 26 Sep. 2011. Television.

Timeless. "Pilot." Season 1, episode 1. Dir. Neil Marshall. MiddKid Productions, 2016. Television.

TNT Guide to Babylon 5. Online video. *YouTube*. 20 Aug. 2014. Web. 26 Jul. 2016. https://www.youtube.com/watch?v=-UZXHVAqhOY

V. Dir. Kenneth Johnson. Warner Bros., 1983. Television movie.

V. Warner Bros., 1984–1985. Television.

V. Warner Bros., 2009–2011. Television.

WarGames. Dir. John Badham. United Artists, 1983. Film.

The West Wing. "Duck and Cover." Season 7, episode 12. Dir. Christopher Misiano. John Wells Productions, 22 Jan. 2006. Television.

Westworld. "The Bicameral Mind." Season 1, episode 10. Dir. Jonathan Nolan. Bad Robot, 4 Dec. 2016. Television.

Index

The Abyss (film) 169
Adams, John 167
The Age of the Pussyfoot 47, 55–56
Agent of Byzantium 156, 188
Alas, Babylon 37
The Alchemist 162
Alias (television series) 56, 57, 181
Aliens (film) 43
All Around the Moon 16
"All Good Bems" 43
All the Myriad Ways 134
"All You Zombies—" 131
Allen, Grant: "Pausodyne" 53
"Alpha Ralpha Boulevard" 146
Alter Ego 2, 21, 22–23, 41, 43, 74, 143, 168
Alternate (or Alternative) History 7, 142, 155–156, 186, 188
Amis, Kingsley 156; New Maps of Hell 101
Anachronism 2, 24–26, 189
Anderson, Poul 128, 129, 156, 171, 188; The Enemy Stars 133, 134; "Time Patrol" 103, 174; "Tomorrow's Children" 37, 100
Andromeda (television series) 89, 90, 91, 103, 111
Antony and Cleopatra 24
Anvil, Christopher 125; Pandora's Legions 125
archetype 22, 74, 119, 156–157, 167, 173
Ariel 21, 65, 162, 187
Artificial Intelligence 3, 7, 11, 27–29, 33, 34, 41, 57, 63, 64, 66, 69, 79, 80, 114, 119, 132
As Far as Thought Can Reach 104
Ashton, Kevin 8, 82, 83
Asimov, Isaac 5, 11, 12, 17, 19, 27, 28–34, 42, 47, 48, 51, 54, 62, 63, 66, 71, 76, 77, 78, 79, 80, 94, 111–114, 119, 120, 121, 122, 123, 126, 129, 143, 144, 147, 154, 160,

170, 171, 173, 178, 179, 181, 184, 188, 190; Asimov on Science Fiction 11, 181, 182, 184; The Bicentennial Man and Other Stories 33, 122; The Caves of Steel 30, 32, 143; The Complete Robot 32, 33, 77, 78; The Foundation Trilogy 144, 170, 173; I, Robot 18, 31, 113, 120, 121; In Memory Yet Green 30; "My Robots" 113; "Opposite" 114; "Reason" 113; The Rest of the Robots 77, 121, 123; Robot Visions 30, 32, 78, 123; Robots and Empire 30, 32, 33; The Robots of Dawn 23, 32, 33
Asimov on Science Fiction 11, 181, 182, 184
"Asylum" 102
Atomics 7, 30, 34–41, 67, 90, 100, 168, 184
Avatar (film) 70, 145
The Avengers (film) 190

Babylon 5 (television series) 31, 139, 173, 185
Babylon 5: A Call to Arms (film) 26
Babylon 5: In the Beginning (film) 139
Bad Code 29, 41–42, 62, 90
Balderston, John L. 138, 141; Berkeley Square 138, 141
Banks, Iain M. 28, 109, 110; Excession 109; The Hydrogen Sonata 110; Look to Windward 110; Matter 110; Surface Detail 110
Barr, Marleen S. 157
Barrat, James 28, 34
Battlestar Galactica (television series) 173
Bear, Greg 90, 171, 172; Blood Music 90; Dinosaur Summer 172
The Beautician and the Beast (film) 180
Beggars in Spain 104
BEM (Bug-Eyed Monster) 7, 42–43
Beneath the Planet of the Apes (film) 167
Benforado, Adam 96

214 Index

Benford, Gregory 44, 111, 115, 171; *Heart of the Comet* 111, 115
Berkeley Square (film) 138
Berkeley Square (play) 138, 141
Bernal, J.D. 128
"Bernie the Faust" 161
Bester, Alfred 132; *Tiger! Tiger!* 132
The Bicentennial Man and Other Stories 33, 122
Bierce, Ambrose Gwinnett 119
Big Dumb Objects 7, 43–44
Blish, James 80, 171; *Jack of Eagles* 132; "Solar Plexus" 80
block universe 138, 139
Blood Music 90
Bobble 7, 45, 52, 142
Bond, James 25, 26, 66, 167, 190
The Borderlands of Science 17, 85, 182
Bostrom, Nick 29
Boulle, Pierre 129; *Planet of the Apes* 129
Bradbury, Ray 6, 20, 129, 170; "Payment in Full" 170; *The Toynbee Convector* 20
Brave New World 149, 187
"Bridesicle" 47, 48
Briggs, Katharine M. 65
Brin, David 1, 2, 29, 43, 72, 77, 79, 82, 87, 111, 114, 124, 129, 144, 145, 148, 152, 153, 155, 160, 171, 179, 184, 188; *Existence* 2, 77; *Foundation's Triumph* 144; *Heart of the Comet* 111, 115; *Heaven's Reach* 146, 153; *Insistence of Vision* 155, 160, 184, 188; *Startide Rising* 43, 145, 147, 153; *Sundiver* 145
Brooks, Rodney 32
Brothers in Arms 54
Brown, Fredric 43, 146; "All Good Bems" 43; "The Star Mouse" 146
Buck Rogers in the 25th Century (television series) 114
Bujold, Lois McMaster 54, 55, 101, 104, 149, 150, 160, 172, 188; *Brothers in Arms* 54; *Cetaganda* 104, 149, 150; *Cryoburn* 55; *Ethan of Athos* 149, 150; "The Mountains of Mourning" 101; *Shards of Honor* 149; *The Vor Game* 101; *The Warrior's Apprentice* 101, 150
"Burning Chrome" 56
Burroughs, Edgar Rice 172, 180, 181
Byron, George Gordon Noel 78, 161, 165; *Manfred* 161, 165

Caliban 21, 22, 65, 168
Calvin and Hobbes 72
Cameron, James 104, 145, 169; *The Abyss* 169; *Avatar* 70, 145; *Dark Angel* 104; *The Terminator* 69, 169, 170, 174; *Terminator 2* 169, 177
Campbell, John W. 30, 45, 102, 123, 129, 171, 185; *Islands of Space* 45
Campbell, Joseph 157, 164, 173, 174; *The Hero with a Thousand Faces* 164, 173
Čapek, Josef 120
Čapek, Karel 5, 78, 120; *R.U.R.* 5, 78, 120
Captain Future 119
Cassandras 8, 157–160, 173, 177
The Castle of the Carpathians 13
The Caves of Steel 30, 32, 143
Cetaganda 104, 149, 150
Chernobyl 39, 40
The China Syndrome (film) 39
The Chronic Argonauts 138
Chronotransference 7, 45–46, 142
Citizen of the Galaxy 178
City 30, 146–147
Clarke, Arthur C. 6, 17, 43, 46, 47, 60, 76, 96, 126, 127, 128, 129, 153, 171, 190; *Earthlight* 84, 115; Hal 9000 60; *Profiles of the Future* 47; *Project Solar Sail* 126, 129; *Rendezvous with Rama* 43; *The Songs of Distant Earth* 95, 150; "The Star" 76; "The Sunjammer" 128; *2001: A Space Odyssey* 153; "The Wind from the Sun" 128
Clarke's Laws 46–47
"The Clock That Went Backward" 137
Clute, John 160, 179, 180, 183; *The Encyclopedia of Fantasy* 160; *The Encyclopedia of Science Fiction* (online) 145; *The Encyclopedia of Science Fiction* (print) 160, 179, 180, 183
Cold as Ice 81, 88, 144, 179
Cole, August 117; *Ghost Fleet* 117
Columbo (television series) 32
A Coming of Age 133
"The Command" 146
The Complete Robot 32, 33, 77, 78
Conan Doyle, Arthur 14, 172, 183; *The Lost World* 172
A Connecticut Yankee in King Arthur's Court 141
Contact 62, 123, 138
Coren, Stanley 154
Cornish, Edward 6
Corpsicle 47–48, 55
Cosmogony of the Future 48–50, 172, 183
Cosmos 128, 179
Coulter, Dauna 128
Countdown (film) 90
Cousteau, Jacques 148

Index 215

Creatures of Light and Darkness 132–133
Cryoburn 55
Cryonics 7, 45, 47, 48, 51–55, 108, 137, 142
Csicsery-Ronay, Istvan 12
Cyberspace 55–57, 58, 73, 111, 152

Dark Angel 104
Darrow, Clarence 79
Data (character) 26, 27, 31, 79, 80, 114, 119, 122, 162, 163
Davies, Paul 58, 73, 138
Davis, Stan 6
"The Day Is Done" 102
"A Death in the House" 96
Death Ray 58–60
de Bergerac, Cyrano 75, 178, 182
de Camp, L. Sprague 141, 146, 156, 188; "The Command" 146; Lest Darkness Fall 141, 156
Deep Blue 7, 29, 60–65, 119, 132
The Defenseless Dead 55
DeGroot, Gerald J. 169
Dekker, Thomas 163, 164
de Lint, Charles 140; "Timeskip" 140
Del Rey 37, 100, 102, 146, 147, 166; "The Day Is Done" 102; "The Faithful" 146, 147; "Nerves" 37, 100
Demolition Man (film) 55
Descartes, Rene 156
The Devil's Elixirs 22
De Vries, Hugo 99
Diamandis, Peter 8, 91
Diaspora 58, 64, 99
Dick, Philip K. 99, 102, 103, 149, 156; "The Golden Man" 102; The Man in the High Castle (novel and Amazon series) 156
Dickson, Gordon R. 104, 152, 153, 172, 178; Wolfling 153, 172, 178
The Discarded Image 162
Disney, Walt 38, 132, 187
Doctor Atomic 167
Doctor Faustus 161, 162, 164
Dr. Strangelove or How I Learned to Stop Worrying and Love the Bomb (film) 68, 170
Doctor Who 43, 96, 114, 137, 141, 143
"Dog Stars" 154
Doomsday Machine 37, 66–71, 112, 125, 167
The Doomsday Men 36, 67
The Door Into Summer 45, 52, 53, 54
Double Star 179
Dowling, David 167

Dozois, Gardner 156
Drexler, K. Eric 105, 106, 107, 129
Durant, Will 79
Dvorsky, George 33, 148, 151
Dyson, George 39, 68

Earth Watch: A Survey of the World from Space 14
Earthlight 84, 115
Edison's Conquest of Mars 15
Egan, Greg 58, 64, 99; Diaspora 58, 64, 99
Einstein, Albert 27, 36, 138, 189
Eiseley, Loren 73, 123, 125, 189; The Immense Journey 189; The Invisible Pyramid 73, 123
Elsie Venner 100, 183
The Emerald Sea 149
Empires in the Dust 188
The Encyclopedia of Fantasy 160
The Encyclopedia of Science Fiction (online) 145
The Encyclopedia of Science Fiction (print) 160, 179, 180, 183
The Enemy Stars 133, 134
Engineer Garin and His Death Ray 58–60
Enriquez, Juan 104, 105, 147
Ethan of Athos 149, 150
Ettinger, Robert 54, 55
Excession 109
Existence 2, 77
Extant (television series) 79, 107, 145

Fairy Tales of the Russians and Other Slavs 176, 185
"The Faithful" 146, 147
The Fantastic Made Visible 20, 161, 182, 188, 190
Fantasy (term) 3, 9, 14, 17, 19, 52, 65, 88, 137, 146, 155, 160–161, 171, 174, 180, 181, 182, 183, 185, 186, 187, 190
Farrand, Phil 134
Faust 7, 67, 77, 78, 161–166
Fermis 66, 71–74, 113, 125
Feynman, Richard 106, 121
Fictions of Nuclear Disaster 37, 68, 100, 166–170
Finnegans Wake 173
"First Contact" 74, 75, 76
First Contact (term) 2, 43, 71, 74–77, 111
Fischer, Bobby 61
Flying to Valhalla 70, 71, 111, 125
Footfall 90,112
Forbidden Planet (film) 30, 31, 37, 43, 100, 119, 152, 167, 168, 187

216 Index

Ford, Paul 26; "The Last Museum" 26
Forde, Pat 46; "In Spirit" 46
The Foundation Trilogy 144, 170, 173
Foundation's Triumph 144
Frank, Pat 37, 67, 167; *Alas, Babylon* 37; *Mr. Adam* 37, 67, 167
Frankenstein 182
Frankenstein Complex 29, 30, 62, 77–80, 120, 161
Freedom's Landing 125
Freezing a Mother-in-Law 53
Freud, Sigmund 21, 22, 143
Friar Bacon and Friar Bungay 163
Friedman, Louis 126
From the Earth to the Moon 16, 17, 115, 127
"From the 'London Times' of 1904" 17–18
Fukushima 39, 40
Futurespeak 34, 47, 104, 157, 183, 184

Gabaldon, Diana 141; *Outlander* 141
Garreau, Joel 100, 105
Gas Giant 80–81, 87
George, Peter 68; *Red Alert* 68
Ghost Fleet 117
Gibbons, Thomas 144; *The Uncanny Valley* 144
Gibson, William 56; "Burning Chrome" 56; *Neuromancer* 56, 111
Gilgamesh the King 178
Girin, Alexander 25
Gleick, James 106, 121
Godspeed 82
Golden, Christie 159; *The Farther Shore* 159
"The Golden Man" 102
GoldenEye (film) 25
Goldenfeld, Nigel 24
Goldilocks Planet 72, 81
Granholm, Jackson 90
Graves, Robert 157, 159
Grazier, Kevin 184
Green, Jonathon 90
Greene, Robert 163; *Friar Bacon and Friar Bungay* 163
Griffith, George 58; *The World Masters* 58
Griffith, Mary 53; *Three Hundred Years Hence* 53
"Grotto of the Dancing Deer" 52
Grumbles from the Grave 185
Gunn, James 71, 112, 123, 125, 170, 171, 186; *The Listeners* 123, 126

Hacking Matter 8
Hal 9000 60

Haley, Andrew 92
Hall, J. Storrs 151, 152
Hamilton, Edmond 119; *Captain Future* 119; *Star Kings* 179
Hamlet (character) 145
Hamlet (play) 81
Hard SF 44, 160, 170–171, 184
Harris, Anthony 163
Have Space Suit—Will Travel 43, 92
Hawking, Stephen 27, 28, 124, 125, 130
Healey, Raymond 37, 100
Heart of the Comet 111, 115
"The Heart of the Serpent" 76
Heaven's Reach 146, 153
Heinlein, Robert A. 17, 43, 45, 52, 53, 54, 55, 92, 93, 115, 131, 142, 178, 179, 185, 186; "All You Zombies—" 131; *Citizen of the Galaxy* 178; *The Door Into Summer* 45, 52, 53, 54; *Double Star* 179; *Grumbles from the Grave* 185; *Have Space Suit—Will Travel* 43, 92; *Methuselah's Children* 53; *The Moon Is a Harsh Mistress* 115; *Red Mars* 186; *Starship Troopers* 43; *Stranger in a Strange Land* 142
Henry IV part I 165
A Heritage of Stars 54
The Hero with a Thousand Faces 164, 173
"He's dead, Jim." 81–82
Heston, Charlton 170
Hindenburg 40
Historical Dictionary of Science Fiction Literature 12, 181
The Hobbit 187
Hoffman, E.T.A. 21, 22, 143; *The Devil's Elixirs* 22; "Mademoiselle de Scudery" 22; "Nutcracker and the King of the Mice" 22; "The Sandman" 143
Holmes, Oliver Wendell 100, 183; *Elsie Venner* 100, 183
Holt, Nathalia 157
Homer 27, 64, 65, 119, 158, 159, 160, 178, 182; *The Iliad* 27, 64, 119, 178; *The Odyssey* 158, 159, 178
Hope, Anthony 179
"How-2" 121, 143–144
Howard, Philip 8, 82
Huxley, Aldous 149, 187; *Brave New World* 149, 187
The Hydrogen Sonata 110

I, Robot 18, 31, 113, 120, 121
Iceman (film) 53
The Iliad 27, 64, 119, 178
The Immense Journey 189
In Memory Yet Green 30

Index 217

"In Spirit" 46
Independence Day (film) 61, 77, 124
Insistence of Vision 155, 160, 184, 188
Internet of Things 8, 9, 82–83, 143, 152
Interstellar (film) 139
The Invisible Boy (film) 31
The Invisible Pyramid 73, 123
Ion Drive 17, 83–86, 87, 130
Island in the Sea of Time 142
The Island of Doctor Moreau 146
Islands of Space 45

Jack of Eagles 132
Jacobsen, Annie 59
Jentsch, Ernst 143, 145
Jonson, Ben 24, 162; *The Alchemist* 162; *Sejanus* 24
A Journey to the Center of the Earth 172
Joyce, James 173; *Finnegans Wake* 173
Jules-Verne, Jean 16
Julius Caesar 24
Jung, Carl 23, 156, 157, 173; *Man and His Symbols* 157, 173
The Jungle Books 154
Juno Spacecraft 7, 86–89
"Just Like Old Times" 45, 46

Kahn, Herman 6, 38, 66, 67, 68
Kaku, Michio 87, 88
Kasparov, Garry 60, 61, 62, 63, 65
Kate and Leopold (film) 142
Kaufman, Harold 17, 84
Kaufman, Scott 38
Kazantsev, Alexander 59, 60
Keegan, Rebecca 170
The Killing Star 70, 71, 80, 81, 111, 112, 171
Kipling, Rudyard 14, 154, 172; *The Jungle Books* 154; "The Man Who Would Be King" 172
Kippenhahn, Rudolf 87
Kludge 29, 41, 62, 89–91, 103
Knowles, Christopher 100
Kramnik, Vladimir 63
Krauss, Lawrence 133, 134, 135
Krell 43, 44, 152, 168, 169
Kress, Nancy 104, 160, 170–171; *Beggars in Spain* 104
Kurzweil, Ray 19, 109; *The Singularity Is Near* 19

"The Lady Who Sailed *The Soul*" 128
Lang, Andrew 52
"The Last Museum" 26
Lawson, Dominic 61
Leeming, David 165, 173

The Left Hand of Darkness 99
Le Guin, Ursula K. 99, 160, 172; *The Left Hand of Darkness* 99
Leinster, Murray 18, 75, 76, 139; "First Contact" 74, 75, 76; "A Logic Named Joe" 18; "Proxima Centauri" 75; "Sidewise in Time" 139
Lest Darkness Fall 141, 156
Levin, Ira 131; *The Stepford Wives* 132
Levy, Steven 61, 65
Lewis, C.S. 162; *The Discarded Image* 162
Ley, Willy 38; "A Station in Space" 38
Lidbetter, Guy 19
Lifehouse 135
"The Limits of Individual Plasticity" 146
Linden, David 90
Lipson, Hod 82
The Listeners 123, 126
Livio, Mario 74, 113
"Locksley Hall" 127
Logan's Run (film) 174
Logan's Run (television series) 119
"A Logic Named Joe" 18
Look to Windward 110
The Lord of the Rings 52, 172
Lost Cities and Vanished Civilizations 188
Lost Colony 171–172, 173, 180
Lost in Space (television series) 122
The Lost World 172
Lost World Story 172–173
Lucas, George 174, 187

MacGyver (character) 90, 91, 92
MacGyver (television series) 91
MacGyver (term) 8, 90, 91–92
"Mademoiselle de Scudery" 22
Maher, Bill 91
Man and His Symbols 157, 173
The Man in the High Castle (novel and Amazon series) 156
"The Man Who Would Be King" 172
"The Man Without a Body" 132
"Mana" 147
Manfred 161, 165
Mann, George 181, 182
Marcus, Gary 90
Marlowe, Christopher 161, 162, 164, 165; *Doctor Faustus* 161, 162, 164
Marooned in Realtime 45
"A Martian Odyssey" 75
Matter 110
McCaffrey, Anne 125; *Freedom's Landing* 125
McCarthy, Wil 9; *Hacking Matter* 8
McIntosh, Will 47, 48; "Bridesicle" 47, 48

McMillan, Priscilla 169
Men Like Gods 183
The Merry Wives of Windsor 26
Metalaw 72, 92–97
Metamorphoses 88
Methuselah's Children 53
Millennium (film) 139, 140
Miller, P. Schuyler 171
Minkowski, Hermann 138
Mr. Adam 37, 67, 167
Mitchell, Edward Page 132, 137; "The Clock That Went Backward" 137; "The Man Without a Body" 132
Monomyth 157, 173–177
The Monomyth in American Science Fiction Films 174, 177
The Moon Is a Harsh Mistress 115
Morevac, Hans 33
Mori, Masahiro 142, 143
Moskowitz, Sam 14, 15
The Mote in God's Eye 43, 126
The Mothman Prophecies (film) 139
"The Mountains of Mourning" 101
"Mozart in Mirrorshades" 118, 136
multiplex parenting 97–99, 130, 151
Munro, John 115; *A Trip to Venus* 115
Musk, Elon 28, 39, 109
Mutant 7, 37, 42, 89, 98, 99–105, 183
"My Robots" 113
The Mysterious Island 92

Nahin, Paul 138
nanotechnology 7, 9, 26, 50, 51, 62, 69, 105–108, 121, 151
Napoleon 44, 155
The Neanderthal Man (film) 23
"Nerves" 37, 100
Neural Lace 29, 108–111, 136
Neuromancer 56, 111
"New Folks' Home" 93, 94
New Maps of Hell 101
Newton, Isaac 27, 37, 115, 162; *A Treatise of the System of the World* 115
Nicholls, Peter 44, 86, 160, 179, 180, 183
Nine Princes in Amber 118
Niven, Larry 43, 55, 90, 126, 129, 134, 136, 171, 172; *All the Myriad Ways* 134; *The Defenseless Dead* 55; *Footfall* 90, 112; *The Mote in God's Eye* 43, 126; *Protector* 172
North, John 87
nuclear fission 34, 37, 40, 67
nuclear fusion 41, 73
"Nutcracker and the King of the Mice" 22

Odd John 104
The Odyssey 158, 159, 178
Okuda, Michael 133, 182
Oltion, Jerry 136; *Mudd in Your Eye* 136
"Omnilingual" 76
"On Fairy-Stories" 146
O'Neill, Timothy 173
Oppenheimer, Robert 169
"Opposite" 114
Out of the Dark 77
Outlander 141
Ovid 88; *Metamorphoses* 88
The Oxford Dictionary of Science Fiction 34, 53, 75, 84, 111, 119, 141, 145, 155, 178
Oxford English Dictionary 3, 21, 24, 27, 34, 53, 54, 56, 66, 80, 81, 84, 90, 92, 99, 105, 111, 114, 119, 123, 126, 127, 132, 136, 141, 160, 180, 181, 182, 183, 184, 190
Oz the Great and Powerful (film) 187
"Ozymandias" 45

Palumbo, Donald 174; *The Monomyth in American Science Fiction Films* 174
Pandora's Legions 125
"Pausodyne" 53
Payne, Robert 24
The Peace War 45
Pellegrino, Charles 70, 71, 76, 80, 111, 112, 125, 171; *Flying to Valhalla* 70, 71, 111, 125; *The Killing Star* 70, 71, 80, 81, 111, 112, 171
Pellegrino, Powell and Asimov's Three Laws of Alien Behavior 71, 76, 111–113
Pemberton, T.E. 53; *Freezing a Mother-in-Law* 53
Person of Interest (television series) 28, 29, 41, 57, 113
Pickover, Clifford 114
Pilkington, Ace G. 2, 37, 100, 159, 161, 176, 185; "Dog Stars" 154; *Fairy Tales of the Russians and Other Slavs* 176, 185; *The Fantastic Made Visible* 20, 161, 182, 188, 190; "Forbidden Planet: Aliens, Monsters and Fictions of Nuclear Disaster" 37, 100; "Introduction: Science Fiction and Fantasy Conquer the World" 161, 190; "Star Trek: American Dream, Myth, and Reality" 159
Piper, H. Beam 76; "Omnilingual" 76
Planet of the Apes (film) 146, 170
Planet of the Apes (novel) 129
Pohl, Frederik 12, 47, 55–56, 64, 66; *The Age of the Pussyfoot* 47, 55–56

Positronic Brain 29, 31, 80, 113–114, 171
Pournelle, Jerry 17; *Footfall* 90, 112; *The Mote in God's Eye* 43, 126
Prehistoric Man 154
Priestley, J.B. 36, 37, 38, 67; *The Doomsday Men* 36, 67
Profiles of the Future 47
Project Chariot 38
Project Plowshare 38, 39
Project Solar Sail 126, 129
Prospero 21, 31, 65, 66, 162, 163, 164, 165, 187
Protector 172
proto-science fiction 21, 27, 48, 75, 178, 182, 187
"Proxima Centauri" 75
Prucher, Jeff 34, 53, 75, 84, 111, 119, 141, 145, 155, 178; *The Oxford Dictionary of Science Fiction* 34, 53, 75, 84, 111, 119, 141, 145, 155, 178

Railgun 8, 114–118
Rank, Otto 173, 175; *The Myth of the Birth of the Hero* 173, 175
Reagan, Ronald 59, 166
Realtime 7, 45, 118–119, 136, 137, 140, 142, 156
Red Mars 186
Reflections 170
Regis, Ed 38, 39, 105
Reichardt, Jasia 142
Rendezvous with Rama 43
"Research Alpha" 103
The Rest of the Robots 77, 121, 123
Rim World 178–179
Ringo, John 149; *The Emerald Sea* 149; *There Will Be Dragons* 149
Robby the Robot 31, 119, 168, 169
Roberts, Adam 120, 181, 182, 187
Roberts, Andrew 44, 155
Roberts, J.M. 12, 19
Robinson, Spider 135; *Lifehouse* 135
Robot 5, 7, 14, 18, 23, 27, 28, 29, 30, 31, 32, 33, 42, 62, 64, 65, 69, 77, 78, 79, 80, 87, 97, 107, 108, 113, 114, 119–123, 131, 132, 142, 143, 144, 145, 146, 147, 150, 151, 152, 166, 170, 178, 179, 186
Robot Visions 30, 32, 78, 123
robotics 34, 79, 123; dictionary 122; laws 29–34, 77, 78, 94, 114, 123
Robots and Empire 30, 32, 33
The Robots of Dawn 23, 32, 33
Rogow, Roberta 34, 47, 104, 157, 183, 184; *Futurespeak* 34, 47, 104, 157, 183, 184
Rohn, Jennifer 182

Romanoff and Juliet (film) 179
Romeo and Juliet 162
Rucker, Rudy 95; "Totem Poles" 95
R.U.R. 5, 78, 120
Ruritania 179–180
Russell, Eric Frank 147; "Mana" 147
Russell, Jeffrey Burton 164; *Witchcraft in the Middle Ages* 164
Rutherford, Ernest 35, 56

Saberhagen, Fred 69, 70; "Starsong" 70; "Without a Thought" 69
Sagan, Carl 62, 123, 127, 128, 138, 179; *Contact* 62, 123, 138; *Cosmos* 128, 179
"The Sandman" 143
Saunders, Russell 129
Sawyer, Robert J. 45, 46; "Just Like Old Times" 45, 46; "You See But You Do Not Observe" 46
Science Fact and Science Fiction 57–58, 161, 163, 165–166, 171, 183, 189–190
science fantasy 180–181
science fiction (term) 181–182
scientific romance 182–184
sci-fi (term) 184–185
Sejanus 24
Sellars, Peter 88; *see also* Doctor Atomic
Separable Soul 185
Serviss, Garrett P. 15, 17, 35; *A Columbus of Space* 15, 35: *Edison's Conquest of Mars* 15
SETI 71, 73, 123–126
Shakespeare, William 21, 24, 37, 51, 65, 81, 88, 145, 152, 165, 168, 178, 179, 182, 187, 189; *Anthony and Cleopatra* 24; *Hamlet* 81, 145; *Henry IV Part I* 165; *Julius Caesar* 24; *The Merry Wives of Windsor* 26; *Romeo and Juliet* 162; *The Tempest* 21, 65, 152, 162, 168, 178, 182, 187, 189; *The Winter's Tale* 165
Shards of Honor 149
Shatner, William 17, 18, 19, 85
Shaw, Bernard 104; *As Far as Thought Can Reach* 104
Sheffield, Charles 7, 14, 17, 43, 54, 55, 81, 82, 88, 129, 144, 171, 179, 182; *The Borderlands of Science* 17, 85, 182; *Cold As Ice* 81, 88, 144, 179; *Earth Watch: A Survey of the World from Space* 14; *Godspeed* 82; *Tomorrow and Tomorrow* 54, 55, 170
Shelley, Mary 13, 160, 182; *Frankenstein* 182
Shelley, Percy Bysshe 45; "Ozymandias" 45

Shumaker, Wayne 162
"Sidewise in Time" 139
Silverberg, Robert 122, 178, 188; *Empires in the Dust* 188; *Gilgamesh the King* 178; *Lost Cities and Vanished Civilizations* 188
Simak, Clifford 30, 52, 54, 55, 93, 94, 96, 112, 121, 133, 134, 135, 143–144, 146–147, 154; *City* 30, 146–147; "A Death in the House" 96; "Grotto of the Dancing Deer" 52; *A Heritage of Stars* 54; "How-2" 121, 143–144; "New Folks' Home" 93, 94; *Prehistoric Man* 154; *Way Station* 52, 93, 133; *Why Call Them Back from Heaven?* 54–55
Singer, P.W. 117
The Singularity Is Near 19
Sirius 146
Slan 103, 104
Smith, Cordwainer 128, 146; "Alpha Ralpha Boulevard" 146; "The Lady Who Sailed The Soul" 128
Smith, P.D. 38, 67, 68
Soddy, Frederick 35; *Wealth, Virtual Wealth, and Debt* 35
"Solar Plexus" 80
Solar Sail 126–130
Solo Parenting 98, 99, 130–131
The Songs of Distant Earth 95, 150
Sontag, Susan 167
speculative fiction 182, 183, 185–188, 190
speculative history 1, 7, 46, 140, 142, 156, 182, 184, 188–189
speculative nonfiction 189–190
Spielberg, Steven 79, 118, 172
Spring Fever 180
Sputnik 38
Stableford, Brian 12, 57–58, 161, 163, 165–166, 171, 181, 183, 189–190; *Historical Dictionary of Science Fiction Literature* 12, 181; *Science Fact and Science Fiction* 57–58, 161, 163, 165–166, 171, 183, 189–190
Stapledon, Olaf 104, 146; *Odd John* 104; *Sirius* 146
"The Star" 76
Star Kings 179
"The Star Mouse" 146
Star Trek (television series) 17, 18, 19, 23, 63, 69, 81, 84, 94, 107, 108, 133, 135, 179, 187
Star Trek (2009 film) 94, 107, 108, 118, 175, 176, 177
Star Trek Beyond 94, 107, 108, 118, 176
Star Trek: First Contact (film) 75, 107, 108, 174

Star Trek Into Darkness (film) 94, 107, 108, 118, 177
Star Trek: The Next Generation (television series) 26, 31, 39, 43, 52, 79, 94, 107, 108, 114, 122, 133, 162, 187
Star Trek II: The Wrath of Kahn (film) 94, 107, 108, 174
Star Trek: Voyager (television series) 94, 107, 108, 140, 144, 158, 159, 173, 176, 187
Star Wars: Episode IV—A New Hope (film) 23, 86, 174, 176–177, 181
Star Wars: Episode VII—The Force Awakens (film) 23, 86, 174, 178, 181
Stargate: Atlantis (television series) 44, 75, 116, 172
Stargate: SG-1 (television series) 44, 76, 91, 133, 136, 172, 177
Starship Troopers 43
"Starsong" 70
Startide Rising 43, 145, 147, 153
"A Station in Space" 38
Stepford Wives (term) 29, 131–132
The Stepford Wives (film) 132
The Stepford Wives (novel) 132
Sterling, Bruce 95, 105, 118, 136; "Mozart in Mirrorshades" 118, 136; "Totem Poles" 95
Stevenson, Robert Louis 21, 22, 23; *The Strange Case of Dr. Jekyll and Mr. Hyde* 21, 22
Stirling, S.M. 142; *Island in the Sea of Time* 142
Stover, Leon 138
The Strange Case of Dr. Jekyll and Mr. Hyde 21, 22
Stranger in a Strange Land 142
Strugatsky, Arkady 76
Strugatsky, Boris 76
Sundiver 145
"The Sunjammer" 128
Superman III (film) 23
Surface Detail 110
Szilard, Leo 36, 38, 67, 68

Taves, Brian 14, 16, 17, 182
Technothriller 25, 26, 91, 190–191
teleportation 2, 7, 110, 132–136, 186
The Tempest 21, 65, 152, 162, 168, 178, 182, 187, 189
Tenn, William 99, 161; "Bernie the Faust" 161; "Venus and the Seven Sexes" 99
Tennyson, Alfred 127; "Locksley Hall" 127
The Terminator 69, 169, 170, 174

Terminator Genisys 175
Terminator 2 169, 177
Terra Nova (television series) 118
There Will Be Dragons 149
Thorne, Kip 138, 139
Three Hundred Years Hence 53
Three-Mile Island 39
Time Bunny 7, 118, 136–137, 142
Time Machine (term) 7, 45, 118, 137–139, 140, 141, 189
The Time Machine (novel) 52, 120, 137, 138, 174
"Time Patrol" 103, 174
Timeless (television series) 141
Timequake 7, 118, 139–140, 141
Timeskip 7, 140
"Timeskip" 140
Timeslip 7, 138, 140–142, 156
Tipler, Frank 135, 150
Tolkien, J.R.R. 52, 146, 160, 174, 187; *The Hobbit* 187; *The Lord of the Rings* 52, 172; "On Fairy-Stories" 146
Tolstoy, Alexei 58–60; *Engineer Garin and His Death Ray* 58–60
Tomorrow and Tomorrow 54, 55, 170
"Tomorrow's Children" 37, 100
Tono-Bungay 36
"Totem Poles" 95
Townes, Charles 59
A Traveller in Time 141, 142
A Treatise of the System of the World 115
A Trip to Venus 115
Tsiolkovsky, Konstantin 17, 48, 49, 84, 127
Turing, Alan 79, 121, 144
Turtledove, Harry 156, 188; *Agent of Byzantium* 156, 188
Twain, Mark 15, 17–18, 141; *A Connecticut Yankee in King Arthur's Court* 141; "From the 'London Times' of 1904" 17–18
Twenty Thousand Leagues Under the Sea 16, 34
2001: A Space Odyssey 153

Uncanny Valley 29, 78, 132, 142–145
The Uncanny Valley 144
The Universe Makers 48, 50, 172, 184
Uplift 145–149, 153
uterine replicator 98, 149–151
utility fog 9, 83, 151–152
Uttley, Alison 141, 142; *A Traveller in Time* 141, 142

V (film) 77
V (television series) 77, 125

Van Vogt, A.E. 102, 103, 104, 170; "Asylum" 102; *Reflections* 170; "Research Alpha" 103; *Slan* 103, 104
Varley, John 139–140; "Air Raid" 140
"Venus and the Seven Sexes" 99
Verne, Jules 13, 14, 16, 17, 34, 35, 48, 49, 81, 92, 105, 114, 115, 127, 137, 169, 171, 172, 181–182, 183, 184, 190; *All Around the Moon* 16; *The Castle of the Carpathians* 13; *From the Earth to the Moon* 16, 17, 115, 127; *A Journey to the Center of the Earth* 172; *The Mysterious Island* 92; *Twenty Thousand Leagues Under the Sea* 16, 34
Vinge, Vernor 33, 45; *Marooned in Realtime* 45; *The Peace War* 45
Von Braun, Wernher 17, 37, 49, 84
Von Hippel, Arthur 105
The Vor Game 101

Waltonen, Karma 157
The War of the Worlds 75
WarGames (film) 56, 69, 166, 170
Warrick, Patricia 18, 30, 121
The Warrior's Apprentice 101, 150
Watson, Sara 143
Watters, Thomas 80
Watterson, Bill 72; *Calvin and Hobbes* 72
Watts, Peter 186
Way Station 52, 93, 133
Wealth, Virtual Wealth, and Debt 35
Weber, David 77, 149, 188; *Out of the Dark* 77
Weinbaum, Stanley 75; "A Martian Odyssey" 75
Wells, H.G. 14, 15, 17, 35, 36, 37, 52, 67, 75, 120, 137, 138, 139, 146, 171, 183, 184, 186; *The Chronic Argonauts* 138; *The Island of Doctor Moreau* 146; "The Limits of Individual Plasticity" 146; *Men Like Gods* 183; *The Time Machine* 52, 120, 137, 138; *Tono-Bungay* 36; *The War of the Worlds* 75; *When the Sleeper Wakes* 52; *The World Set Free/The Last War* 35, 36, 67
Wendelstein 7-X (W7X) stellerator 41
The West Wing (television series) 39
Westworld (television series) 42
When the Sleeper Wakes 52
Why Call Them Back from Heaven? 54–55
Wilson, Taylor 41
"The Wind from the Sun" 128
The Winter's Tale 165
Witchcraft in the Middle Ages 164
"Without a Thought" 69

Wodehouse, P.G. 180; *Spring Fever* 180
Wolfling (novel) 153, 172, 178
Wolfling (term) 147, 152–154, 173, 179
Wollheim, Donald 48, 50, 51, 172, 184; *The Universe Makers* 48, 50, 172, 184
Woodman, David 162, 163
The World Masters 58
The World Set Free/The Last War 35, 36, 67

Yefremov, Ivan 76; "The Heart of the Serpent" 76

"You See But You Do Not Observe" 46

Zahn, Timothy 133, 171; *A Coming of Age* 133
Zarkadakis, George 11, 28, 66
Zebrowski, George 70, 71, 80, 111, 112, 113, 171; *The Killing Star* 70, 80, 111, 112, 171
Zelazny, Roger 118, 132–133; *Creatures of Light and Darkness* 132–133; *Nine Princes in Amber* 118
Zipes, Jack 12, 20
Zubrin, Robert 126

www.ingramcontent.com/pod-product-compliance
Ingram Content Group UK Ltd.
Pitfield, Milton Keynes, MK11 3LW, UK
UKHW041949140426
5217IPUK00014B/710